PERGAMON INTERNATIONAL LIBRARY
of Science, Technology, Engineering and Social Studies
*The 1000-volume original paperback library in aid of education,
industrial training and the enjoyment of leisure*

Publisher: Robert Maxwell, M.C.

ELECTRONICS—
from Theory into Practice

APPLIED ELECTRICITY AND ELECTRONICS DIVISION

General Editor: P. HAMMOND

SOME OTHER BOOKS IN THIS SERIES

ABRAHAMS, J. R. & PRIDHAM, G. J.
Semiconductor Circuits: Theory, Design and Experiments

BADEN FULLER, A. J.
Microwaves

BROOKES, A. M. P.
Basic Instrumentation for Engineers and Physicists

CRANE, P. W.
Electronics for Technicians

GATLAND, H. B.
Electronic Engineering Applications of Two-port Networks

HAMMOND, P.
Electromagnetism for Engineers

HARRIS, D. J. & ROBSON, P. N.
The Physical Basis of Electronics

HOWSON, D. P.
Mathematics for Electrical Circuit Analysis

PRIDHAM, G. J.
Solid State Circuits

SPARKS, J. J.
Junction Transistors

WHITFIELD, J. F.
Electrical Installations Technology

The terms of our inspection copy service apply to all the above books. Full details of all books listed will gladly be sent upon request.

UHI
Millennium
Institute

Please return/renew this item by the last date shown

Tillibh/ath-chlaraidh seo ron cheann-latha mu dheireadh

PERGAMON PRESS

OXFORD · NEW YORK · TORONTO

SYDNEY · PARIS · FRANKFURT

U.K.	Pergamon Press Ltd., Headington Hill Hall, Oxford OX3 0BW, England
U.S.A.	Pergamon Press Inc., Maxwell House, Fairview Park, Elmsford, New York 10523, U.S.A.
CANADA	Pergamon of Canada Ltd., Box 9600, Don Mills M3C 2T9, Ontario, Canada
AUSTRALIA	Pergamon Press (Aust.) Pty. Ltd., 19a Boundary Street, Rushcutters Bay, N.S.W. 2011, Australia
FRANCE	Pergamon Press SARL, 24 rue des Ecoles, 75240 Paris, Cedex 05, France
WEST GERMANY	Pergamon Press GmbH, 6242 Kronberg/Taunus, Pferdstrasse 1, Frankfurt-am-Main, West Germany

First edition 1966

Second edition 1976

Library of Congress Cataloging in Publication Data

Fisher, Jack Edward.
Electronics from theory into practice.

(Applied electricity & electronics division)
(Pergamon international library)
Bibliography: p.
Includes index.
1. Electronics. I. Gatland, Howard Bruce, joint author. II. Title.
TK7815.F58 1976   621.381   75-44016
ISBN 0-08-019857-0 Vols 1 & 2
0 08 019855 4 Vol 1
0 08 019856 2 Vol 2

Printed in Great Britain by A. Wheaton & Co., Exeter

Contents of Volume 1

Contents of Volume 2

Preface to Volume 1

To all students of electronics there comes a time when a specification is presented to them, and they are expected to turn their theoretical knowledge into practice. Many find this a difficult step to take. The aim of this book is, where possible, to formalize design procedures covering a wide range of electronic circuitry, and thus to bridge the gap between theory and practice. It is also hoped that the book will be of use to practising engineers, particularly those trained in other disciplines who, due to the widespread application of industrial control and automation, are obliged to undertake a certain amount of electronic design.

Besides being available complete, in one hard-cover volume, the book has been produced in two flexi-cover volumes for the convenience of students. The first two chapters of Volume 1 introduce the reader to the bipolar and field-effect transistor, the unijunction transistor and the silicon-controlled rectifier, and show how data sheets, provided by the manufacturer, are used in design calculations. Also included are a number of devices which, by belonging to the realm of microwaves, are outside the scope of this book. This has been done for reference purposes. A third chapter traces the development of integrated circuits and gives details of the characteristics of such which are currently in use. It ends with an introduction to charge-coupled devices.

The remaining two chapters of this volume are devoted to amplifier design. Each contains a brief treatment of theory, limited to the extraction of necessary design relationships. Design procedures are established, followed by worked design examples to meet given specifications. In conclusion, two appendices give simple introductions to the Laplace Transform and to Network Analysis, in explanation of methods which are occasionally employed in the text of this book.

Because it deals with devices and fundamental concepts, it is hoped that this first volume will be suitable for first-year courses in electronic design. It leads naturally into a second volume which treats operational amplifiers, power supplies, oscillators and digital techniques in a similar manner.

Cranfield   J. E. FISHER

Design Examples

CHAPTER 1

The Semiconductor

INTRODUCTION

A semiconductor material is one having a specific resistance inter-mediate between that of an insulator and a conductor, the value of which increases rapidly with rising temperature. Considering the atomic structure of such material, if sufficient energy is provided, by heating for instance, electrons will be released from their nuclei, each leaving behind it a hole. Under the influence of an electric field, an electric current will flow, which may be regarded as a movement of electrons in one direction and a movement of holes in the opposite direction. In the case of a pure or *intrinsic* semiconductor the numbers of holes and free electrons are always equal. The two materials which have been most commonly used are germanium and silicon, both of which come from chemical Group IV. Gallium arsenide, however, is a material which is now being used in ever increasing fields of application.

If a semiconductor is doped with an element from Group V, say arsenic, the equality of free electrons and holes will no longer exist, there being an excess of free electrons. An electric current through such a material will then consist mostly of a flow of electrons in one direction and relatively few holes moving in the opposite direction. In this case the electrons are called *majority carriers* and the holes, *minority carriers*. A semiconductor doped in this way is known as n-type material since the majority carriers possess negative charge.

A similar state of affairs will occur if the semiconductor is doped with an element such as indium from Group III. However, in this case an excess of holes will exist and these are the majority carriers. Since the majority carrier possesses positive charge such a material is known as p type. In the production of semiconductor devices, it is often required that the level of doping be controlled. The more heavily doped a material is, the lower is its resistivity. Heavily doped material is identified by the symbols n^+ and p^+.

1

1.1. The junction diode[1]

If a piece of semiconductor material is doped with p-type impurity at one end and n-type impurity at the other, then there will exist a junction between the two types. Some holes in the p region will diffuse into the n region leaving the p region slightly negative. Similarly, electrons from the n region will diffuse into the p region leaving the n region slightly positive. In a layer between the n and p regions, holes and electrons recombine and, since this layer is now depleted of free charge carriers, it is called the *depletion layer*. This layer acts as a potential barrier which opposes any further diffusion of charge, and the junction assumes a state of dynamic equilibrium. The condition is illustrated in Fig. 1.1a.

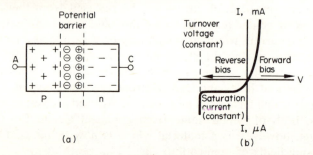

Fig. 1.1. (a) Semiconductor junction showing the potential barrier caused by the diffusion of charge carriers. Under these conditions a potential exists between A and C. (b) Characteristic curve of a semiconductor diode. Note the change in current scale as the curve passes through the origin.

If bias is applied to the terminals such that A is positive with respect to C, it has the effect of reducing the thickness of the depletion layer. The potential barrier is thus reduced and current will flow. This current increases exponentially with increasing voltage until the potential barrier is reduced to zero, when it is limited only by the resistance of the semiconductor material. If the bias is now reversed, the potential barrier is increased and the majority carrier is blocked. There is, however, a finite current which flows, called the *reverse saturation current*. As the reverse bias is increased this current remains constant until the turnover point is reached, when the current increases rapidly at constant voltage (Fig. 1.1b).

Thus, if a junction is biased in the forward direction, a fairly large current will flow, but under reverse bias conditions, provided the turnover voltage is not reached, the current is extremely small. In other words the device acts as a rectifier.

Applications. Typical applications of the semiconductor diode are illustrated in Fig. 1.2.

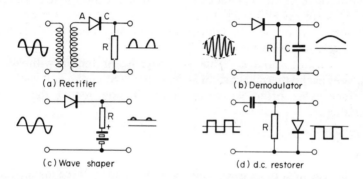

(a) Rectifier   (b) Demodulator

(c) Wave shaper   (d) d.c. restorer

Fig. 1.2. Some applications of the semiconductor diode.

1.2. Leakage current

This current varies from 1 to 200 μA for small germanium units, while silicon units have values much less than this. In both cases the current is dependent on temperature as is shown in Fig. 1.3 for a transistor with open-circuit emitter.

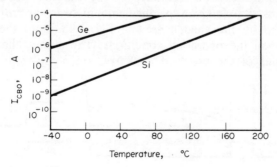

Fig. 1.3. Relationship between reverse saturation current and temperature for germanium and silicon.

The main causes of leakage current are:

(a) *Minority carriers* generated thermally, *radiation effects* and *crystal imperfections*. In these cases hole–electron pairs are formed and the charge carrier, which is the minority carrier, travels across the junction which is biased against the majority carrier. Generally, thermal generation is the more significant cause and the reverse saturation current will increase by 10 per cent for 1°C rise (or doubles for an 8°C rise). Because of this, the use of germanium devices is restricted to temperatures below about 70°C. Silicon, however, is usable up to 150°C.

(b) *Surface leakage.* This is a significant factor with silicon devices as the thermal current is very small within the unit itself, and surface paths, often caused by contamination, reduce the reverse resistance.

1.3.   Diode transient response

The transient response of diodes is modified by three main effects.

Carrier storage.   When a diode is conducting there are minority carriers in the cathode region. For a *pn* diode, these carriers are holes injected from the *p* anode into the lightly doped *n*-type cathode. On reversal of the applied voltage, the minority carriers in the cathode region return to the anode. The reverse current will fall to the normal leakage value only when this process is complete. The peak value of reverse current is determined by the magnitude of the applied reverse voltage and the series resistance in the circuit. The carrier storage can be characterized by the time taken for the reverse current to fall by a specified amount, or by the *stored charge*, represented by the shaded area in Fig. 1.4a. Both these values depend upon the operating conditions. The stored charge varies from values of the order of 10,000 pC (pico-coulombs) for low-

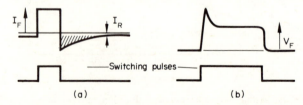

Fig. 1.4. Diode transient response. The shaded area in (a) represents stored charge.

frequency types to 100 pC for high-frequency types. Gallium arsenide diodes are the best in this respect since devices are available with the total stored charge of the order of a few pico-coulombs.

Turn-on transient. When a diode is "turned on", initially the voltage across the device is higher than the steady value, as is illustrated in Fig. 1.4b. The initial value peaks to approximately twice the steady value but returns to that voltage in less than 0.1 μs.

Junction capacitance. The capacitance across a reverse biased junction is a function of the reverse voltage applied to it. In junction diodes, the depletion layer capacitance is inversely proportional to the square root of this applied voltage. The range of capacitance, for small reverse voltages, is from approximately 1 pF for high-speed devices to 20 pF for general-purposes devices. The effect is made use of, for the manufacture of small capacitors for inclusion in integrated circuits. An ability to vary the junction capacitance is the essential feature of the *varactor diode* described in § 1.6.

1.4. Diode logic

By employing planar techniques a number of diodes can be formed on the same substrate, forming a multiple diode unit. Such devices have the advantage of small physical size and similarity of diode characteristics, and can be used in networks to provide logical operations.

OR gate. In Fig. 1.5a, when the inputs are at zero volts, if the forward voltage drop across the diodes is ignored, the output voltage is zero. In this quiescent condition the current in R is $I_k = V_k/R$ and each of the input sources must provide a current $I_k/3$.

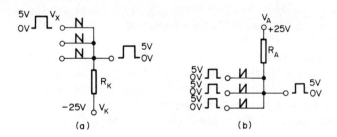

Fig. 1.5. Diode logic arrangements. (a) OR gate, (b) AND gate, for positive logic.

If any of the inputs is raised to $+5$ V the output rises to the same voltage. The network thus functions as an *OR* gate and is described by the Boolean equation $f(x) = A + B + C$ in which the symbol $+$ is read as OR. In the active state the diodes with anodes at 0 V are reverse biased, while the conducting diodes pass a total current $I_k = (V_k + V_x)/R$.

AND gate.   If any of the inputs of Fig. 1.5b are at 0 V they draw sufficient current to ensure that the output will remain at zero. The output will rise to the upper level only when all the inputs are at the upper voltage level. The network thus performs as an *AND* gate and is described by the Boolean function $f(x) = A.B.C$. This is read as A *and* B *and* C. If all the inputs are at 5 V, the output will be 5 V. If now one of the inputs is taken to 0 V the output will return to 0 V also.

Both the gates described are said to have *positive logic* since the outputs are functions of the high levels of input signals. If the output signals are considered with reference to the low levels of input signals the gates will obey *negative logic*. In this form the AND gate performs as an OR gate, and the OR gate provides the AND function.

1.5.   Functional survey of diode types

Junction diodes.   Reference to Fig. 1.21 shows that, when planar construction is used, the exposed junction is protected by the oxide layer and contamination is prevented. As a result, silicon planar devices can be mass produced with stable characteristics. Because of their low leakage current they are generally preferred to germanium types, particularly in switching networks and for high-speed computing purposes. However, the forward voltage necessary for conduction is of the order of 0.5 V for silicon and 0.15 V for germanium, so for large currents the latter will be more efficient. In addition to germanium and silicon, a semiconductor material of significance is gallium arsenide. Diodes made with this material can be used in temperatures up to several hundred degrees centigrade and are employed for high-speed operation.

Point contact diodes.   The earliest form of semiconductor device was the point contact diode, consisting of a tungsten spring on n-type germanium. In this, passage of a small current forms a

p-type alloy at the point of contact. Because of the small contact area the forward resistance is higher than for junction types. However, the point contact diode has excellent high-frequency performance and was preferred for high-speed switching. For high-voltage operation, lightly doped germanium is required, this having greater resistivity, but the result is an inferior high-frequency performance. By using a gold wire instead of tungsten, a diode with low forward resistance but good high-frequency performance is obtained. Such a device is known as a *gold-bonded* diode.

Reference diodes. Silicon diodes, when forward biased, provide a reasonably well-defined voltage of the order of 0.6 V. The mechanics of breakdown which occur when a junction is reverse biased are discussed in § 1.10. The breakdown characteristic can be used as a voltage reference. By using heavily doped material a zener breakdown can be introduced at predetermined voltages, up to about 15 V. Higher voltage-range reference diodes make use of avalanche breakdown.

Schottky diodes.[2] The structure of a Schottky diode is illustrated in Fig. 1.6a, rectification taking place at the metal-to-semiconductor interface. The device functions on the same principle

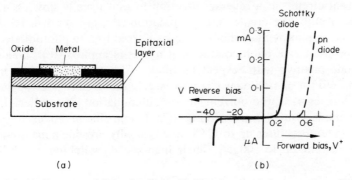

(a) (b)

Fig. 1.6. (a) Structure of Schottky diode. (b) The characteristic curve shows that its forward voltage drop is half that of a normal diode.

as early "cat's whisker" radio detectors which employed metal contacts on silicon–carbide, but modern methods of production enable large areas of metal–semiconductor contact to be achieved. Diode action does not involve minority carriers, current flow being attributed only to majority carriers, the electrons which exist in the conduction bands of the materials. The elimination of minority carriers means that the problem of stored charge, mentioned earlier, is effectively eliminated. This accounts for the extremely fast switching capability of the device. The other important feature is that it has a forward voltage drop which is approximately half that of a *pn* junction diode, thus making it more efficient when used in low-voltage high-current power supplies. The characteristic curve of a typical Schottky barrier diode is given in Fig. 1.6b, in which reverse current is scaled in μA and forward current in mA.

Photodiodes.[3]   The two processes to be considered here are *photoconductive* and *photovoltaic*. When light falls on a photoconductive material, light energy is absorbed and electron–hole pairs are created. If an external field is now applied, holes move in one direction, electrons move in the other, and an electric current results. The higher the density of illumination, the greater is the current.

Photovoltaic devices differ in that a voltage is generated across a cell when light shines upon it. They normally consist of *pn* junctions and the voltage appears across a junction with the *p* material positive. This voltage is capable of causing a current to flow in an external circuit, in a reverse direction to that usually flowing in a diode. Again, the greater the illumination, the greater will be the current. A photovoltaic diode is now referred to as a photodiode. It can be used as a photoconductor if a bias is applied to it, making cathode positive with respect to anode. Although germanium has been used for photodiodes it is now largely replaced by silicon since the *dark current*, the current at zero illumination, is smaller. The photodiode symbol is given in Fig. 1.7a. Planar silicon diodes may be used in temperatures up to 125°C and typically provide a maximum-light current up to 500 μA. Their high-speed switching capability

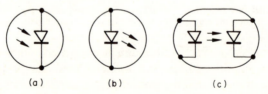

(a)   (b)   (c)

Fig. 1.7. Opto-electronic symbols, (a) photodiode, (b) light-emitting diode and (c) optically coupled isolator.

make them suitable for use in fast tape readers and in character-recognition equipment.

Light-emitting diodes.[4] The light produced by an LED is due to electroluminescence, a process which converts electrical energy into light energy. It depends upon the fact that electromagnetic radiation can be produced when a physical system relaxes from a state of higher energy to one of lower energy. The principle of action is illustrated in Fig. 1.8. Associated with a *pn* junction are two

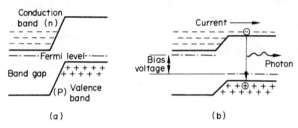

Fig. 1.8. The mechanism of photo emission. Light energy is given up when an electron moves from a high-energy level to one of low energy.

energy levels, the *conduction* band and the *valence* band, separated by a band gap in which no free electrons can exist. With no applied bias the *Fermi level*, which describes the population of the conduction and valence bands, has a value which is constant throughout the structure, and the inherent voltage across the junction is approximately equal to the band gap. The application of a forward bias reduces the potential barrier, and current flows. In general this is comprised of a movement of both holes and electrons, but of these it is the electron current which is the more significant. As electrons move from the *n* layer to the *p* layer, they are moving from one energy level to another. In the *p* region they combine with holes and their extra energy is given up. With a suitable bandgap, this released energy is in the form of visible light. The light emission from an LED increases with current. The device is used with a lens to ensure the correct spatial distribution of light, and the colour is determined by the semiconductor material used. Materials currently in use include gallium arsenide, yielding infrared radiation, gallium arsenide phosphide, red or orange, and gallium phosphide which can emit yellow or green light. The eye has maximum sensitivity to green light so this latter material has a greater overall efficiency than the others. The symbol used for the LED is shown in Fig. 1.7b. The forward and reverse characteristic is similar to that of any other *pn* junction, the forward voltage drop varying from 1 to 4 V, with reverse breakdown voltages in the range 5 to 50 V.

Optically coupled isolators. An OCI consists of a photoemissive device and a photosensitive device integrated into a single unit. It is intended for use to transfer a signal from input to output where it is necessary for these to be isolated from each other. A typical unit employs a gallium-arsenide light-emitting diode with either a silicon photodiode or a phototransistor, and provides high-voltage insulation between input and output terminals. The device symbol for the diode–diode unit is shown in Fig. 1.7c.

1.6.   R.F. and microwave diodes[5]

Although the application of these devices is, in general, outside the scope of this book they are included here for the sake of references.

PIN switching diodes.[6] These are silicon junction devices which are used to switch and control radio-frequency signals. A PIN diode consists of a layer of intrinsic silicon, the *i* layer, sandwiched between heavily doped *n*- and *p*-type material. When a d.c. reverse bias is applied, the carriers in the *i* layer are swept out and only a small leakage current flows through this depleted region. The application of a forward bias reduces the potential barriers at the *pi* and *in* junctions, so holes from the *p* layer and electrons from the *n* layer enter the *i* region. Here they neutralize each other, so no space charge will exist to limit current flow. Thus, the *i* layer of a PIN diode is almost devoid of carriers when reverse bias is applied, and filled with carriers when forward biased. It can therefore function as a switch.

In practice an oscillating signal is superimposed upon a bias current, and the signal is controlled by varying the bias. In the forward bias condition the device acts as a pure resistance to r.f. By varying the bias, this effective resistance can be varied, and the r.f. signal modulated. In the reverse bias condition the effective impedance is capacitive. PIN diodes are being used as switches at frequencies in the range 1.5 MHz to 40 GHz and power which can be handled may be as much as a megawatt. Applications include aerial switching and duplexing, phase shifting, band changing and modulation.

Varactor diodes and parametric amplification.[7] As the reverse bias across a *pn* junction is increased the depletion layer gets thicker. This is analogous to separating the plates of a parallel-plate capacitor, and the capacitance falls. A device which exhibits a voltage-dependent junction capacitance is known as a varactor diode (see Fig. 1.9a). These have found wide use in the field of

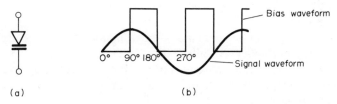

Fig. 1.9. The varactor diode and its use for parametric amplification.

parametric amplification, in which a bias waveform and a signal are applied to a junction simultaneously. Let the applied bias waveform be a square wave, of twice the frequency of the signal waveform, and having the phase relationship shown in Fig. 1.9b. At 90° the signal is introducing a maximum charge into the junction capacitance. At this point a step increase in bias occurs and the capacitance of the device becomes smaller. Since $Q = CV$ and the charge is an instantaneous constant, the signal potential, developed across the diode, increases. The continuing signal swing from 90° to 180°, where it is zero, is therefore greater in amplitude than it otherwise would have been. At this point the bias waveform returns to zero and the capacitance is restored to its former value. However, the rate of change of signal voltage with time, dV/dt, is maximum at 180° and considerable signal overshoot results. At 270° the process is repeated and a second maximum signal voltage is achieved in the opposite direction, following the first *pumped up* maximum. Due to the work done by the pumping pulse, as the process continues each half cycle sees more signal charge stored and greater signal voltage peaks obtained. A state of equilibrium is reached when the losses inherent in moving charges to and fro balance out further voltage magnification. Voltage gains of 1000 or more may be achieved using this technique. In practice the varactor diode would obtain its signal from a source such as a tuned circuit. It is likewise necessary for the *pump* to supply its signal from a current source. Although the foregoing explanation has used a square wave for the pumping signal, this can be replaced by a sinewave of similar amplitude and phase. As a result, however, signal amplification will be reduced.

Varactor diodes and harmonic generation.[8] When a sinusoidal signal acts on a non-linear impedance the signal becomes distorted. Fourier analysis of the distorted signal then reveals the presence of harmonics of the fundamental frequency. Since the variable reactance (varactor) diode has a depletion layer capacitance which varies with applied voltage in a non-linear manner, it may be used as a frequency multiplier. An alternating voltage is applied across the diode, which generates harmonics of the applied frequency. The use of filters and suitable resonators then enables power at the desired harmonic frequency to be extracted. The performance of the device is dominated by the nature of the *pn* junction. The effective depletion layer capacitance is given by

$$C_{dl} = C_0/[1 + (V/\phi)]^m$$

where C_0 is the capacitance with zero bias, V is the applied bias, ϕ is the barrier potential and m depends upon the junction doping profile. Typical values of ϕ are 0.5 V for silicon and 1.1 V for gallium arsenide. The value of m varies from 0.5 for an abrupt junction to 0.3 for a linear graded one. Although higher harmonics are generated, the varactor is usually employed for doubling, trebling or quadrupling an applied frequency.

Step recovery diodes.[9] The SRD, which is a variation of the varactor diode, is used for frequency multiplication by six and upwards. One single unit can thus replace a chain of varactor multipliers. A specially selected junction doping profile makes it possible for an applied r.f. signal to switch the SRD rapidly between a low-capacitance reverse biased state and a relatively high-capacitance forward-biased state. The capacitance discharge, each cycle, into an inductance, produces a train of impulses which are rich in harmonics. These impulses are used to shock excite a resonator, having a loaded Q of $n\pi/2$ and this in turn produces a damped ringing waveform at n times the input frequency. A schematic diagram of such an arrangement is given in Fig. 1.10.

The impatt diode[10] is a source of microwave power capable of

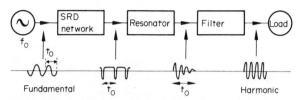

Fig. 1.10. Schematic diagram to illustrate the use of the step recovery diode to generate harmonics of a fundamental frequency.

producing 1 W c.w. at 50 GHz and up to 50 W of pulse power at 10 GHz. Its name is derived from its mechanics of operation, i.e. impact avalanche and transit time. The structure of the device is given in Fig. 1.11 together with a simplified equivalent circuit of the diode mounted in a typical package.

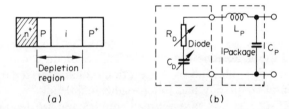

Fig. 1.11. Structure of an impatt diode and its equivalent circuit when mounted in a typical package.

The essential feature of an oscillator is that it converts energy from an applied steady state into an alternating state. Inevitably, some transient is necessary to cause oscillations to commence and then, providing that certain conditions are met in terms of gain and phase relationships, oscillations can be maintained. In the impatt diode oscillator, an applied reverse bias provides the input d.c. energy. As this bias is increased the resulting field clears the p and i regions of carriers thus creating a depletion layer. As a result there remains a high electric field at the abrupt n^+p junction. With increasing bias some critical value of field is reached (about 350 kV/cm), when avalanche breakdown occurs (see § 1.10), and electron-hole pairs are generated. The electrons enter the n^+ region and may be ignored. The change of holes, however, enter the depletion region and drift across to the p^+ region, thus causing a current to flow in some external circuit. For an electric field greater than 5 kV/cm the drift through the depleted silicon region is at constant velocity. In operation the applied reverse bias is of such value that the abrupt junction field is just below the critical avalanche value. Now let some transient cause oscillations to commence. If these cause the field to oscillate above and below this critical value, *bunches* of holes will traverse the diode, one for each half cycle of oscillation, and cause pulses of current to flow. If the width of the depletion layer, and therefore the transit time, is correct, these pulses will have the right phase relationship to ensure that oscillations are maintained.

The trapatt diode is a device which makes use of the same principle, in which the avalanche breakdown of an abrupt junction gives rise to bunches of holes whose transit time through the unit is defined. In simple terms, an impatt structure radiates into some microwave circuit. If the circuit presents a short circuit to these oscillations, power is reflected back into the diode. If the correct phase relationships are obtained this results in very large voltage swings in the avalanche region. The impatt diode can be made to oscillate in this mode, and such operation is characterized by a lower fundamental frequency and much higher efficiency.

Gunn effect devices[11] are not really diodes, but are dealt with here for the sake of completeness. They are generally made from gallium arsenide, this being one of the materials known as *two-valley* semiconductors. In such material the conduction band has two energy levels which can be occupied, a low-energy level in which electrons have low effective mass and high mobility, and a high-energy level in which they have high effective mass and low mobility. A gunn effect device consists of a slice of n-type gallium arsenide with ohmic contacts at each end. In the absence of an applied bias nearly all the electrons occupy the low-energy band. With an increasing applied bias a point is reached when the electrons jump into the high-energy band and hence suffer a reduction in their mobility. The situation now is that electrons from the bias source are arriving at the cathode at one speed, but moving away through the semiconductor at a lower speed. The result is an accumulation of charge at the cathode and this *domain*, as it is called, grows until it effectively neutralizes the field at the contact and causes it to fall below the critical level for energy band transfer. The accumulation of charge ceases and the domain travels through the semiconductor, at a speed determined by the applied field, and in the form of a sharp spike. When the domain has left the semiconductor, the field at the cathode rises again and the process repeats. The effective frequency is therefore determined by the domain velocity and the length of the device. For good efficiency the unit is operated in a resonant circuit and, typically, would produce 200 mW at 10 GHz.

Tunnel diodes[12] have been used to generate frequencies in excess of 100 GHz but due to their poor power-handling capabilities they have been largely overshadowed by impatt and gunn devices. Thus, future applications will probably be limited to low noise microwave amplifiers and high-speed logic elements.

It has been shown that when a *pn* junction is formed, a potential barrier is created and that, in order for an electron to cross this barrier, it must be given extra energy equal to that of the potential barrier. However, early experimenters discovered that, if the depletion layer is very thin, electrons could pass from one side of the barrier to the other with less energy than was apparently necessary. They attributed this phenomenon to a tunnelling effect. The process is illustrated in Fig. 1.12. The doping level of the device is of a

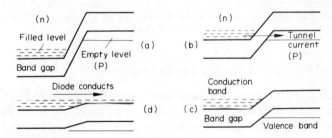

Fig. 1.12. Tunnel diode action. Tunnelling occurs when the filled level of the conduction band is opposite the empty level of the valence band.

magnitude which partially fills the conduction energy band with electrons and leaves a lot of unfilled levels in the valence band. Figure 1.12a represents the condition for zero bias. As the applied bias is increased, conduction band electrons are given more energy and the level is raised, relative to that of the valence band. With increasing bias (Fig. 1.12b) filled levels of the conduction band come opposite empty levels of the valence band and a tunnelling current flows from one to the other. Having reached a peak value, this current decreases as the filled level passes the empty level, and further increase in bias then causes current to flow with normal diode action. The result is a *negative resistance* effect in the characteristic curve of Fig. 1.13a. For comparison, the characteristic curve of a typical *pn* rectifying diode is given in Fig. 1.13b. It will be noticed that for the tunnel diode, zener breakdown occurs almost immediately as reverse bias is applied. Any device which exhibits a negative resistance characteristic may, by the addition of suitable circuitry, be used as an oscillator.

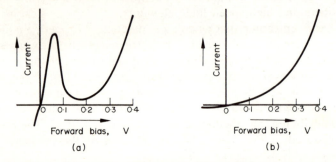

Fig. 1.13. The characteristic curve of (a) the tunnel diode is compared with that of (b) a normal *pn* diode.

1.7. The junction transistor[13]

A typical alloy-type junction transistor is illustrated in Fig. 1.14a and consists of a sandwich of doped semiconductor material, the base being more lightly doped than the collector or emitter, and very thin. As the names suggest, the emitter roughly corresponds to the cathode of a thermionic valve, and the collector to the anode (see Appendix C). A *pnp* transistor is one in which the emitter and collector are *p*-type material and the base is *n* type. There is a *pn* junction between emitter and base, and an *np* junction between base and collector, so the device may be represented by two diodes as in Fig. 1.14b. With the batteries connected as shown, the emitter–base junction is forward biased, thus reducing the potential barrier, while the base–collector junction is reverse biased and its potential barrier therefore increased, as illustrated in Fig. 1.15. The forward bias of

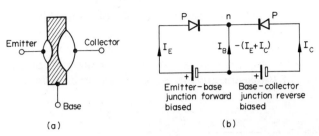

Fig. 1.14. (a) Schematic representation of an alloy transistor. (b) Representation of a *pnp* transistor by two diodes. I_E is the emitter current, I_C is the collector current and the base current $I_B = -(I_E + I_C)$.

the emitter–base junction causes holes to be injected into the base where, since the base is n type, they are minority carriers. Provided that they do not recombine with electrons in the base region, these holes will diffuse towards the base–collector junction. This junction is, however, forward biased for such minority carriers, which are swept into the collector region and give rise to a current in the collector circuit.

Not all the holes flowing from the emitter into the base will pass into the collector region, as a small proportion will combine with electrons in the n-type base. This loss of charge in the base layer is made good by a flow of base current. Varying the base current varies the voltage across the emitter junction, and so controls the emitter–collector current.

The action of an *npn* transistor is similar, except that the minority carriers through the base region are electrons instead of holes. Whichever type of transistor is used, it is necessary that the emitter–base junction be forward biased, that is, with the supply

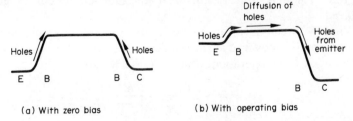

Fig. 1.15. Potential diagram of a *pnp* transistor showing the effect of biasing. The potential barriers are shown as hills which the holes have to surmount.

negative connected to device *n* and the supply positive connected to device *p*. Likewise, the base–collector junction must have reverse bias. This means that a *pnp* transistor has its collector taken to negative while an *npn* type has its collector taken to positive. The two types of transistor are therefore represented diagrammatically as in Fig. 1.16.

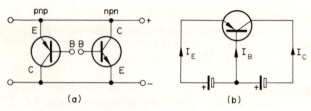

Fig. 1.16. (a) Diagrammatic representation of *pnp* and *npn* transistors. (b) Fundamental current relationships in a junction transistor.

1.8. Fundamental current relationships

In Fig. 1.16b,

$$I_E + I_B + I_C = 0. \tag{1.1}$$

Let

$$\Delta I_B = i_b,$$
$$\Delta I_C = i_c,$$
$$\Delta I_E = i_e.$$

Let the symbol α denote the ratio between collector current and emitter current thus, $\alpha = -i_c/i_e$. This will be less than unity, since i_c is less than i_e because of recombination of the minority carriers in the base region. A typical value for α (the short-circuit current gain) is 0.98.

When a circuit is arranged such that the signal enters at the emitter and is recovered at the collector, the transistor is said to be in *common base* operation. Since $-i_b = i_e + i_c$,

$$\frac{i_c}{i_b} = \frac{i_c}{-(i_e + i_c)} = \frac{\alpha}{1 - \alpha}.$$

The ratio between collector current and base current is denoted by the symbol β:

$$\beta = \alpha/(1-\alpha) \quad (\beta = 49 \quad \text{for} \quad \alpha = 0.98).$$

Usually β ranges from 10 to 150. This implies a current gain if the signal current enters at the base and is recovered at the collector. A transistor used in this way is said to be in *common emitter* operation. This is the most frequently used mode of operation, because of the current gain which may be achieved in this way.

In common emitter operation, the leakage current I_{CEO} that flows (when $I_E = 0$) is approximately βI_{CBO}. That is, β times the leakage current in common base operation. As the collector current is small, and current gain is a function of collector current (low for small currents), the effective β will be of the order of half the usual value. Thus, in the case where $\alpha = 0.98$ and $\beta = 49$, if $I_{CBO} = 5 \ \mu$A, then $I_{CEO} \doteqdot 125 \ \mu$A.

1.9. Elementary considerations of frequency effects[14]

Charge carriers take a finite time to diffuse through the base and this causes a reduction in current gain at high frequencies. If the current gain at zero frequency $\alpha_0 = i_c/i_e$, then

$$\alpha = \alpha_0 \Big/ \left(1 + \frac{jf}{f_\alpha}\right) \qquad (1.2)$$

with reasonable accuracy, where f_α is the frequency at which the gain falls to $0.7\alpha_0$. Similarly, since $\beta = \alpha/(1-\alpha)$,

$$\beta = \beta_0 \Big/ \left(1 + \frac{jf}{f_\beta}\right), \qquad (1.3)$$

where β falls to $0.7\beta_0$ at $f_\beta = (1-\alpha)f_\alpha$. This is illustrated for the 2N 3715 in Fig. 1.17.

The curve for the short-circuit current gain follows closely the locus of a passive lag (a capacitor shunted by a resistor), up to a frequency of $f_\alpha/2$. This relationship is used by Ebers and Moll in switching analysis.[15] The common emitter short-circuit current gain β can be obtained from Fig. 1.17b. It falls from β_0 to $0.7\beta_0$ at approximately $(1-\alpha)f_\alpha$, and to unity at $\alpha_R = 0.5$. This latter frequency is known as f_1:

$$f_1 = \frac{f_\alpha}{1 + \phi}, \qquad (1.4)$$

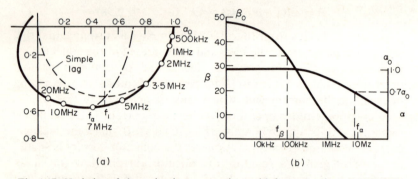

Fig. 1.17. Variation of short-circuit current gain α with frequency for a 2N 3715. Up to 3.5 MHz the curve for α closely follows that of a passive lag.

where $\phi = 0.2$ for homogeneous base transistors and varies between 0.2 and unity for graded base transistors (§ 1.12).

Gain-bandwidth product f_T [14]

If β is plotted against frequency, both on logarithmic scales, the relationship is as shown in Fig. 1.18a. The gain–bandwidth product f_T is obtained by projecting the straight section of the curve to intercept the unity gain line (0 dB). It can thus be determined by

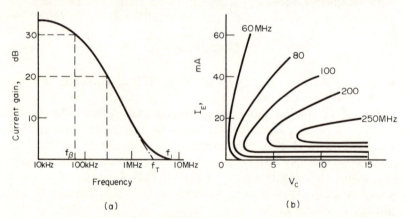

Fig. 1.18. (a) Curve showing relationship between f_β, f_T and f_1. For $f_T = 3$ MHz and $\beta = 10$ (20 dB), band width $= f_T/\beta = 300$ kHz. (b) Variation in gain–bandwidth product, with operating conditions, for a high-frequency transistor.

finding β at any frequency on the 20 dB/decade slope and multiplying this value by the frequency at which it is measured. Thus, in Fig. 1.18a, $f_T = 3$ MHz, and at 20 dB ($\beta = 10$) the bandwidth is 300 kHz. The gain–bandwidth product varies with operating conditions, as is illustrated in Fig. 1.18b.

1.10. Voltage breakdown[16]

When the reverse voltage applied to a junction is increased above a certain figure a "breakdown" will occur and the reverse resistance will fall from the usual high value to a low one. This, in itself, will not damage the unit but it is possible, under the breakdown condition, for large currents to flow if there is no controlling external resistance.

The three mechanisms of breakdown are:
(a) Punch through.
(b) Zener.
(c) Avalanche.

(a) *Punch through*

The width of the depletion layer of a reverse biased junction is approximately proportional to the square root of the voltage, and it is possible for it to extend to the emitter junction. This provides a way for holes to reach the collector, independent of the base, and causes a short circuit between emitter and collector. Thin-based transistors such as surface barrier types have voltage limitations due to punch through, typically having punch-through values of $V_{PT} = -5$ V.

(b) *Zener breakdown*

This occurs when the number of added impurities is high (the greater the doping level the lower the resistivity). Application of reverse bias can cause the valence band of the p material to overlap the conduction band of the n material, as in Fig. 1.12 for the tunnel diode, and the junction will break down. This is usually of no significance in transistors having low doping levels, but the phenomenon is usefully employed in Zener and tunnel diodes. Zener breakdown is usually confined to below 15 V.

(c) *Avalanche*

In lightly doped junctions, carriers entering the depletion layer are accelerated by the field set up by the collector–base voltage. If this voltage is high enough, the carriers can ionize fixed atoms giving rise to hole–electron pairs which, in turn, can cause further ionization. The voltage at which this happens is the "avalanche voltage".

The maximum collector voltage rating is frequently low for applications where the base resistance is high, and is a minimum for open-circuited base V_{CEM}. The maximum collector rating is V_{CBM} which is the collector–base voltage with emitter open circuit. In pulse circuits, the base–emitter junction is frequently reverse biased, and there is similarly a maximum reverse voltage V_{BEM}.

Typical values for a transistor could be:

$$V_{CEM} = 40 \text{ V}; \quad V_{CBM} = 20 \text{ V}; \quad V_{BEM} = 6 \text{ V}.$$

In most applications a safe working voltage will be between V_{CEM} and V_{CBM}.

In Fig. 1.19a is plotted a typical collector characteristic with V_C extended into the breakdown region. Figure 1.19b illustrates how the maximum peak collector voltage varies with base resistance.

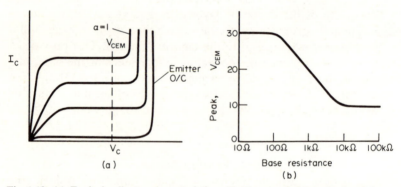

Fig. 1.19. (a) Typical collector characteristic, with V_C extended into the breakdown region, for common base operation. (b) Maximum permissible collector voltage, for a given base resistance, with transistor in common emitter connection.

1.11.   Power dissipation[17]

In a transistor most power dissipation occurs in the collector–base junction where the voltage gradient is a maximum. Using peak instantaneous values, collector dissipation,

$$P_C \doteq v_{CE} i_C \doteq \frac{T_{JM} - T}{\theta},\qquad(1.5)$$

where T_{JM} is the maximum junction temperature, T is the ambient temperature, and θ is the rise in temperature for a given unit of power dissipation (thermal resistance).

Example. Ambient temperature 35°C, $T_{JM} = 70°C$ and $\theta = 1°C/mW$,

$$P_C = \frac{35°C}{1°C/mW} = 35 \text{ mW}.$$

Figure 1.20 illustrates how the use of a heat sink reduces the value of θ for a given transistor.

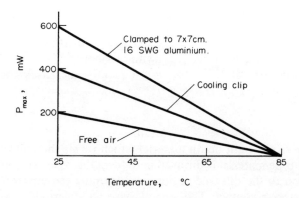

Fig. 1.20. The use of a heat sink reduces the thermal resistance θ, and thus increases the maximum permissible dissipation for a given transistor.

1.12. Summary of transistor types

Low and medium frequency types

Two general processes have been employed: (a) alloy and (b) grown junction. Alloy types have a low collector resistance when bottomed and are suitable for amplifiers and switches at low and high powers. However, f_1 is limited to approximately 20 MHz and beyond this frequency base thinness leads to mechanical weakness.

Medium to high-frequency types

Most new techniques are developing devices in this category, providing values of f_1 between 20 and 1000 MHz.

Surface barrier

Historically, the first mass-produced h.f. transistor, the surface barrier type, uses an accurately controlled etching process to produce a very thin base upon which the emitter and collector are electrodeposited. The main limitation is low punch-through voltage (6 V), and this manufacturing technique has been superseded by the micro-alloy process.

Diffused base

These are sometimes known as Graded Base or Drift Transistors. By varying the resistivity of the base material, an electric field is formed which causes the charge carriers to drift across the base, rather than diffuse across, as with a homogeneous base. This reduces the transit time and consequently increases the upper frequency limit.

Mesa

The device is built up on material which forms the collector, and opposite type material is diffused in to form the base. Two strips are alloyed on to the diffused region, one forming the emitter and the other the base contact. The narrow base width provides a high f_1, and the size of the collector allows high-power dissipation.

Epitaxial mesa

The disadvantage of the mesa transistor is the large carrier storage which limits switching speed, and the high saturation resistance. The latter disadvantage occurs because of the need to have high resistivity material accompanied by a thickness which is adequate for mechanical strength. Both the limitations are overcome in the epitaxial mesa, by depositing a thin, high resistivity epitaxial layer on a relatively thick low resistivity substrate, as illustrated in Fig. 1.21.

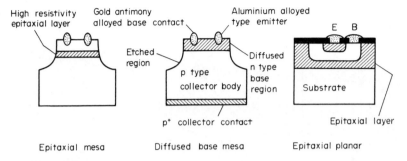

Fig. 1.21. Construction of three different transistor types. Compared with the alloy device of Fig. 1.14, the larger collector volume permits much greater dissipation.

Planar

Using silicon as the substrate, the surface is passivated by oxidization, and diffusion is carried out through etched areas. Such a technique reduces leakage current and provides high gain at low current levels. Because of surface passivation the shoulders do not need to be etched away, as is done in the mesa transistor. The advent of the silicon epitaxial planar technology made possible the mass production of improved and reliable, stable devices, and these types now dominate the transistor field.

1.13. Static characteristics of the junction transistor

In normal operation the input junction of a transistor is forward biased and power has to be supplied to the input terminals. The output current is consequently a function of both input current and voltage.

Input characteristics

In Fig. 1.22 curves are plotted of base current against base voltage, with collector voltage constant, for the common emitter mode of operation.

Let the differential input resistance be denoted by the symbol h_i, such that

$$h_i \underset{\text{def}}{=} \left[\frac{\delta V_B}{\delta I_B}\right] \quad (V_C \text{ constant}). \tag{1.6}$$

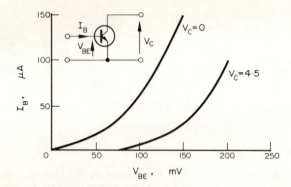

Fig. 1.22. The input characteristic of a transistor, from which the differential input resistance h_i is obtained.

Then on the input characteristic curve, the slope at any working point is equal to the reciprocal of h_i. It is apparent that the input curves are a function of the collector voltage. The input resistance does not change greatly with collector voltage but varies widely with base voltage. Typical operating values of h_i are in the range 500 to 2000 Ω for small units.

Transfer characteristic

This relates output current to input and is linear except for very small and very large currents (Fig. 1.23). The slope of the transfer

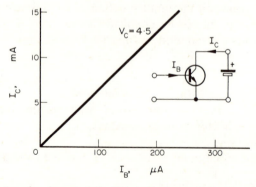

Fig. 1.23. The transfer characteristic. This is linear except for very small and very large currents. The slope of the graph is h_f, the current gain.

characteristic is the current gain of the transistor which, for the common emitter case, is defined as

$$h_f \underset{\text{def}}{=} \left[\frac{\delta I_C}{\delta I_B} \right] \quad (V_C \text{ constant}), \text{ previously denoted by } \beta. \quad (1.7)$$

Both the input and transfer characteristics can be easily measured in the laboratory by supplying the input current from a constant current generator.

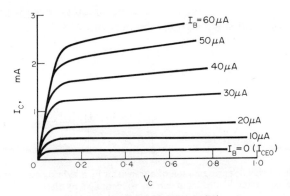

Fig. 1.24. The output characteristic.

The output characteristic

A family of curves is drawn (Fig. 1.24) relating collector current to collector voltage with base current as parameter. The slope of the curves is the output conductance defined as:

$$h_0 \underset{\text{def}}{=} \left[\frac{\delta I_C}{\delta V_C} \right] \quad (I_B \text{ constant}). \quad (1.8)$$

It is commonly of the order of 50×10^{-6} S (the Siemen S = amps/volts).

Voltage feedback characteristic

The fourth relationship is between output and input voltage and is the voltage feedback characteristic. The slope of this curve is:

$$h_r \underset{\text{def}}{=} \left[\frac{\delta V_B}{\delta V_C}\right] \quad (I_B \text{ constant}). \tag{1.9}$$

Voltage feedback ratios are typically of the order of 5×10^{-4}.

1.14. Small signal representation

Hybrid equivalent circuit

This representation makes use of h parameters. For given operating conditions and small signals, the network of Fig. 1.25 represents

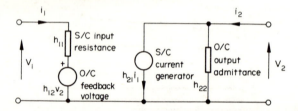

Fig. 1.25. Small signal representation of a transistor by a hybrid equivalent network. The voltage generator in the input circuit is identified by the + sign. A current generator is employed in the output circuit.

a transistor at low frequency. The equations of this network are:

$$v_1 = h_{11}i_1 + h_{12}v_2, \tag{1.10}$$
$$i_2 = h_{21}i_1 + h_{22}v_2. \tag{1.11}$$

And, in matrix form,

$$\begin{bmatrix} v_1 \\ i_2 \end{bmatrix} = \begin{bmatrix} h_{11} & h_{12} \\ h_{21} & h_{22} \end{bmatrix} \begin{bmatrix} i_1 \\ v_2 \end{bmatrix}, \tag{1.12}$$

where
(a)

$$h_{11} = h_i = \left[\frac{\delta V_{\text{in}}}{\delta I_{\text{in}}}\right] \quad (V_C \text{ constant}),$$

$$= \frac{v_1}{i_1} \quad \text{when } v_2 \text{ is zero}, \tag{1.13}$$

that is when the output is short circuited for signal currents;

(b)

$$h_{12} = h_r = \left[\frac{\delta V_{in}}{\delta V_{out}}\right] \quad (I_B \text{ constant}), \tag{1.14}$$

$$= \frac{v_1}{v_2} \quad \text{when } i_1 \text{ is zero,}$$

that is when the input is open circuited;

(c)

$$h_{21} = h_f = \left[\frac{\delta I_{out}}{\delta I_{in}}\right] \quad (V_C \text{ constant}), \tag{1.15}$$

$$= \frac{i_2}{i_1} \quad \text{when } v_2 \text{ is zero,}$$

that is when the output is short circuited;

(d)

$$h_{22} = h_0 = \left[\frac{\delta I_{out}}{\delta V_{out}}\right] \quad (I_B \text{ constant}), \tag{1.16}$$

$$= \frac{i_2}{v_2} \quad \text{when } i_1 \text{ is zero,}$$

that is when the input is open circuited.

The values of these h parameters will vary according to the mode of operation of the transistor. To denote which mode is in use a second subscript, e, b or c, is added to the h symbol. Thus,

h_{ie} = Input impedance, common emitter.
h_{rb} = Voltage feedback ratio, common base.
h_{oc} = Output admittance, common collector.
h_{fe} = Forward current transfer ratio, common emitter.

Hybrid parameters vary with the operating conditions, particularly the collector current, as shown in Fig. 1.26. The short-circuit current gain, h_{fe}, remains relatively constant and can be used as a "figure of merit" when comparing devices.

The short-circuit input resistance, h_{ie}, is inversely proportional to collector current. An estimate of this parameter is provided by the relation

$$h_{ie} \doteqdot r_e h_{fe},$$

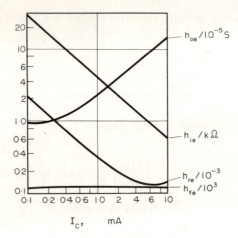

Fig. 1.26. Variations of hybrid parameters with changes in collector current. Note that h_{ie} is inversely proportional to collector current, and h_{fe} remains almost constant.

where r_e is the incremental resistance of the forward-biased diode curve shown in Fig. 1.1b, which can be expressed as

$$I = I_{co} \exp\left(-\frac{eV}{kT}\right),$$

where I_{co} is the reverse saturation current, k is Boltzmann's constant, e the electron charge, and T the absolute temperature. At normal temperature

$$\frac{kT}{e} = 25 \text{ mV}.$$

Differentiating the above expression with respect to the diode current

$$\frac{dI}{dV} = I\bigg/\frac{kT}{e}.$$

The forward resistance is

$$r_e = \frac{dV}{dI} = \left(\frac{kT}{e}\right)\bigg/I = 25 \times 10^{-3}/I.$$

Thus r_e is 25 Ω if $I_E = 1$ mA.

An approximation for the short-circuit input resistance is

$$h_{ie} = \frac{25h_{fe}}{I(\text{mA})}\ \Omega.$$

1.15. Transistor biasing[17]

For operation, the emitter–base junction must be forward biased and base current will flow. Other devices, such as the field effect transistor or the thermionic valve, normally have the input element reverse biased and, ideally, zero current flows.

Base current biasing

Referring to Fig. 1.27,

$$I_B = \frac{V_{CC} - V_{BE}}{R_B}. \tag{1.17}$$

Since V_{BE} is very much less than V_{CC}, $I_B \doteqdot V_{CC}/R_B$. Thus, for $R_L = 1\ \text{k}\Omega$ and $V_{CC} = 8\ \text{V}$, for a bias current of $60\ \mu\text{A}$,

$$R_B = \frac{V_{CC}}{I_B} = \frac{8\ \text{V}}{60\ \mu\text{A}} = 133\ \text{k}\Omega.$$

$I_C = BI_B + I_{CEO}$, where B is the large signal current gain.

$$B = \frac{I_C - I_{CEO}}{I_B} = 57 \quad \text{in this case.} \tag{1.18}$$

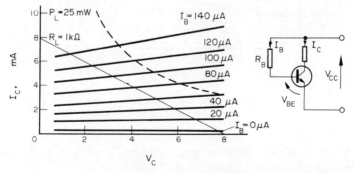

Fig. 1.27. Determination of the necessary bias current for a given operating condition. Note that the operating point must lie below the curve indicating maximum permissible collector dissipation.

The small signal current gain β, evaluated at the working point, is 52, showing that the error involved in using β instead of B is negligible.

Advantages of current biasing
1. Simplicity and small number of components required.
2. Biasing is largely independent of V_{BE}, the base–emitter voltage. The V_{BE} required to maintain a given collector current falls with increasing temperature. If V_{BE} is very much less than V_{CC} (or V_{BB}) the bias current I_B is largely independent of V_{BE}.

Disadvantages
1. I_{CEO} is not controlled and can limit the collector swing or even bottom the transistor in extreme conditions. (I_{CEO} is doubled for each 80°C rise in temperature.)
2. The current gain increases with temperature and can vary greatly with different transistors. Since the standing current $I_C = \beta I_B + I_{CEO}$, the working point can also differ. This is illustrated in Fig. 1.28.

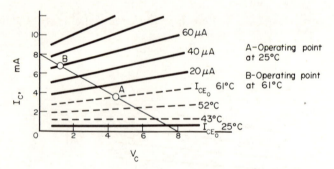

Fig. 1.28. Change in operating point due to increasing temperature, for a germanium transistor with base current biasing.

In spite of the disadvantages, current biasing is very useful, especially for experimental work. The change in operating point is greatly reduced if large collector currents are used (so that I_{CEO} is only a small fraction of I_C) and the loads are small.

Collector feedback biasing

Ideally, the transistor should be biased by a method which prevents excessive shift of the working point due to temperature changes. Indeed in the absence of such stabilization, or where the stabilization is insufficient, thermal runaway may result. Thermal runaway occurs when an increase in temperature causes an increase in collector current, which in turn results in a further increase in temperature. The effect is cumulative and can result in the destruction of the transistor, although it normally occurs only in high-power stages.

The circuit of Fig. 1.29a is the simplest method of providing a

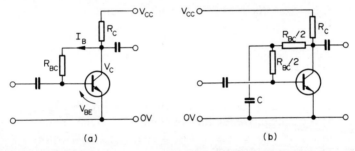

Fig. 1.29. (a) Stabilization of the operating point using collector feedback biasing. (b) The feedback resistor is decoupled to eliminate loss in gain due to negative feedback.

degree of stabilization to the operating point. The performance equations of this circuit are:

$V_C \doteq V_{CC} - I_C R_C$ (neglecting I_B in comparison with I_C),

$I_C = \beta I_B + I_{CEO}$,

$I_B \doteq V_C / R_{BC}$ (neglecting V_{BE} in comparison with V_C).

Combining these equations,

$$I_C \doteq \frac{\beta V_{CC}}{R_{BC} + \beta R_C} + \frac{I_{CEO} \cdot R_{BC}}{R_{BC} + \beta R_C}. \tag{1.19}$$

Differentiating with respect to the leakage current I_{CEO},

$$\frac{\delta I_C}{\delta I_{CEO}} = \frac{1}{1 + \beta R_C / R_{BC}}. \tag{1.20}$$

That is, a change δI_{CEO} causes a change of $(1/K)\delta I_C$, where

$$\frac{1}{K} = \frac{1}{1 + \beta R_C / R_{BC}}. \tag{1.21}$$

The factor K is the stabilization factor and should obviously be as large as possible. In this case, this is achieved by using a transistor with a large β.

Calculation of R_{BC}

If as in the previous example,

$$\beta = 52, \quad R_C = 1\,k\Omega, \quad I_B = 60\,\mu A \quad \text{and} \quad V_C = 4.5\,V,$$
$$R_{BC} = V_C / I_B = 4.5\,V/60\,\mu A = 75\,k\Omega.$$

This provides a stabilization factor $K = 1 + \beta R_C / R_{BC} = 1.7$.

Disadvantages
1. The collector current must always be greater than I_{CEO}.
2. There is negative feedback at signal frequency unless decoupling is employed, as in Fig. 1.29b.
3. The degree of stabilization is relatively small.

The method is applicable for amplifiers in which overall signal inversion occurs. It has the advantage that only one capacitor is used and consequently introduces only one additional time constant. This is a very important consideration in the case of feedback amplifiers. The form of biasing can be applied over more than one stage, for amplifiers that have overall signal inversion, as shown in Fig. 1.30.

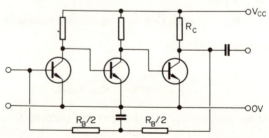

Fig. 1.30. A three-stage amplifier with feedback stabilization. If the collector voltage of the output transistor is half V_{CC}, the stabilization factor is $K = 2$.

Emitter resistor stabilization

This method is generally used with a voltage dividing network to provide the base voltage. Referring to the circuit of Fig. 1.31, the base voltage is set by the voltage divider consisting of R_1 and R_2:

$$I_E = V_E/R_E = (V_B - V_{BE})/R_E \doteq V_B/R_E,$$

where V_B is much greater than V_{BE} (V_B is usually of the order of $10 V_{BE}$).

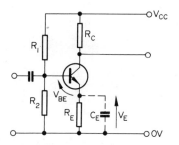

Fig. 1.31. Emitter resistor stabilization. The capacitor C_E decouples R_E to prevent loss in gain due to negative feedback.

If R_1 and R_2 are very small and R_E is very large, the circuit is indistinguishable from the common base configuration. The leakage current is then I_{CBO} (not I_{CEO}) and of the order of $10 \mu A$, and $I_C = I_E + I_{CO}$. Over a wide temperature range α only changes by a small percentage and hence a good degree of stabilization is obtained.

The circuit may be represented by Fig. 1.32, in which

$$R_B = R_1 R_2/(R_1 + R_2).$$

Then, providing that R_C and R_E are not very large,

$$I_C \doteq I_{CEO} + \beta I_B, \tag{1.22}$$

$$I_E + I_C + I_B = 0, \tag{1.1}$$

$$I_B = \frac{V_{BB} - (V_{BE} - R_E I_E)}{R_B}. \tag{1.23}$$

From these three equations,

$$I_B = \frac{V_{BB}}{R_B + R_E} - \frac{V_{BE} + R_E I_C}{R_B + R_E}, \tag{1.24}$$

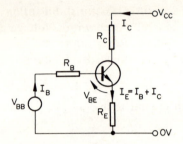

Fig. 1.32. Equivalent representation for determining the stabilization factor.

and

$$I_C = \frac{I_{CEO}}{K} + \frac{\beta V_{BB}}{R_E + R_B} \cdot \frac{1}{K}, \tag{1.25}$$

if V_{BE} is much less than V_E, and where

$$K = 1 + \frac{\beta R_E}{R_E + R_B}. \tag{1.26}$$

Differentiating I_C with respect to I_{CEO},

$$\frac{\delta I_C}{\delta I_{CEO}} = \frac{1}{K}. \tag{1.27}$$

K is thus the stabilization factor and, if R_E is very much greater than R_B, approximates to $1 + \beta$. The designer can thus select his degree of stabilization by suitable choice of values for R_E and R_B.

DESIGN EXAMPLE 1.1

Required, a peak output signal of 3 V without distortion. From the collector characteristic, let the operating point chosen to suit this requirement be $V_C = 8$ V and $I_C = 3$ mA, for a transistor operating from a supply of 12 V. The collector voltage swing is from 5 to 11 V, so a maximum of 5 V is available for V_E, which should be somewhat less than this figure. Let V_E be 3 V so that, for $I_C = 3$ mA, $R_E = 1$ kΩ.

The ratio R_B/R_E is chosen according to the degree of stabilization required. If $R_B/R_E = 3$,

$$R_B = \frac{R_1 R_2}{R_1 + R_2} = 3 \text{ kΩ}. \tag{1.28}$$

As I_B is usually small, $V_B \doteq V_{CC} \cdot R_2/(R_1 + R_2)$. But $V_B = V_{BE} + V_E$ and since V_{BE} is much less than V_E,

$$V_B = \frac{R_2}{R_1 + R_2} \cdot V_{CC} \doteq V_E. \qquad (1.29)$$

Therefore

$$\frac{12 R_2}{R_1 + R_2} = 3 \quad \text{and} \quad R_1 = 3R_2.$$

Substituting in eqn. (1.29), $R_2 = 4\,\text{k}\Omega$ and $R_1 = 12\,\text{k}\Omega$. Make them $4.7\,\text{k}\Omega$ and $12\,\text{k}\Omega$, which are preferred values. To give the required V_C, $R_C = 4\,\text{V}/3\,\text{mA} \doteq 1.2\,\text{k}\Omega$, and the circuit is as shown in Fig. 1.33.

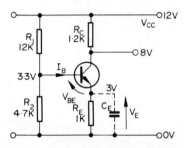

Fig. 1.33. Circuit of Design Example 1.1.

A smaller value of R_B would increase the stability, but would:

1. draw extra current from the power supply,
2. reduce the amplifier input resistance and provide a lower resistance shunt path for input signal current.

For most a.c. applications R_E is decoupled with a capacitor to eliminate negative feedback. R_E, in effect, increases the effective input resistance from h_{ie} to $h_{ie} + h_{fe}R_E$ thereby reducing the voltage gain.

1.16. Transistor amplifier characteristics[18]

The three modes of operation are illustrated in Fig. 1.34a. The signal diagrams represent signal paths, with all the direct voltage supplies replaced by short circuits. The transistor can be represented by the standard hybrid network of Fig. 1.25, in which case

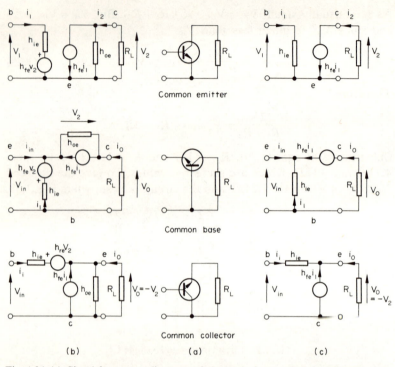

Fig. 1.34 (a) Signal frequency diagrams of the basic forms of transistor operation. The supplies are considered to be short circuited. (b) The hybrid networks for each connection. The common emitter elements are rearranged for the other two connections. (c) Approximate networks when R_L is small.

Table 1.1.

	Common emitter	Common base	Common collector
h_{11}	h_{ie}	$h_{ib} = \dfrac{h_{ie}}{1 - h_{re} + h_{fe} + \Delta h_e}$	$h_{ic} = h_{ie}$
h_{12}	h_{re}	$h_{rb} = \dfrac{\Delta h_e - h_{re}}{1 - h_{re} + h_{fe} + \Delta h_e}$	$h_{rc} = 1 - h_{re}$
h_{21}	h_{fe}	$h_{fb} = \dfrac{-(h_{fe} + \Delta h_e)}{1 - h_{re} + h_{fe} + \Delta h_e}$	$h_{fc} = -(1 + h_{fe})$
h_{22}	h_{oe}	$h_{ob} = \dfrac{h_{oe}}{1 - h_{re} + h_{fe} + \Delta h_e}$	$h_{oc} = h_{oe}$

the *parameters* will be different for each connection. (See Table 1.1 where the elements of the other connections are given in terms of the common emitter parameters.)

Alternatively, the common emitter hybrid equivalent network can be rearranged for each connection as in Fig. 1.34b. If R_L is less than $1/h_{oe}$, the networks can be considerably simplified without introducing much error (Fig. 1.34c).

Current gain

The current gain for the hybrid network is, from Fig. 1.35b,

$$\frac{i_2}{i_1} = \frac{h_{21}G_L}{h_{22}+G_L}. \tag{1.30}$$

If G_L is large compared with h_{22}, the current gain approaches the short-circuit value h_{21}. For the common emitter connection,

$$h_{21} = \beta = h_{fe}.$$

For common base, from Fig. 1.34c,

$$i_{in} = -i_1 - h_{fe}i_1,$$
$$i_o = h_{fe}i_1,$$

and the current gain, h_{fb}, is

$$\frac{i_o}{i_{in}} = \frac{-h_{fe}}{1+h_{fe}}.$$

If $h_{fe} = 49$, the current gain is -0.98 for common base operation, and numerically equal to α. The negative sign signifies that current flowing into the emitter causes current to flow out of the collector, opposite to the positive direction of flow, which by convention is considered as flowing into a device (see Fig. 1.25).

The current gain for the common collector connection is also negative. From Fig. 1.34c, equating the currents at the upper node,

$$i_0 = -i_1 - h_{fe}i_1,$$

whence the current gain is

$$\frac{i_0}{i_1} = -(1 + h_{fe}).$$

Thus for $h_{fe} = 49$, the common collector short-circuit current gain is

$$h_{fc} = -50.$$

Equation (1.30) can be used to find how the current gain varies with load resistance.

The current gain falls to half the short-circuit value when the load conductance is equal to h_{22}. As h_{oe} is typically $20\,\text{k}\Omega$ and h_{ob} is $1\,\text{M}\Omega$, it is apparent that the common emitter current gain will fall off at lower values of load resistance, than for common base. This is shown in Fig. 1.35.

(a)

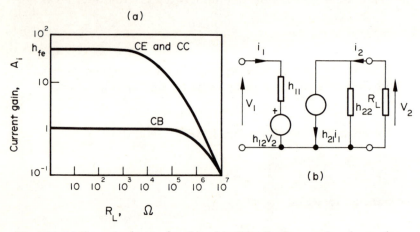

Fig. 1.35. (a) Current gain as a function of load for the three modes of operation. Note the low value for common base. (b) Hybrid equivalent network which is the same for all modes of operation.

Voltage gain

From Fig. 1.35b, and application of Ohm's Law,

$$v_2 = \frac{-h_{21}}{h_{22} + G_L} i_1 = \frac{-h_{21}}{h_{22} + G_L} \cdot \frac{v_1 - h_{12}v_2}{h_{11}}.$$

Thus the voltage gain,

$$\frac{v_2}{v_1} = \frac{-h_{21}}{h_{11}h_{22} - h_{12}h_{21} + h_{11}G_L} = \frac{-h_{21}}{\Delta h + h_{11}G_L}, \tag{1.31}$$

where $\Delta h = h_{11}h_{22} - h_{12}h_{21}$.

In many practical examples, G_L is much greater than $\Delta h / h_{11}$ and

$$\frac{v_2}{v_1} \doteq \frac{-h_{21}}{h_{11} G_L}. \tag{1.32}$$

Thus for the common emitter connection, using the simplified networks of Fig. 1.34c, by KCL (Appendix A),

$$v_0 = (h_{fe} + 1) R_L i_1, \quad \text{the voltage gain}$$

$$\frac{v_0}{v_{\text{in}}} \doteq \frac{v_o}{h_{ie} i_1 + v_o} = \frac{(1 + h_{fe}) R_L}{h_{ie} + (1 + h_{fe}) R_L}. \tag{1.33}$$

If R_L is greater than $h_{ie} / (1 + h_{fe})$, the voltage gain of the common collector is close to unity.

Because of this, and as the signal is not inverted, the connection is referred to as the *emitter follower*.

The more precise expressions for voltage gain can be calculated from the networks of Fig. 1.34b. All are functions of load resistance, and are plotted with respect to R_L in Fig. 1.36.

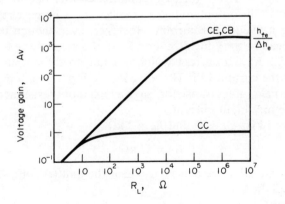

Fig. 1.36. Voltage gain as a function of load resistance. The common collector gain is slightly less than unity.

Common emitter input resistance

Referring to the simplified hybrid network of Fig. 1.34c, the input resistance is obviously h_{ie}, and will be $2\,k\Omega$, approximately, for a small transistor operating at 1 mA emitter current.

Common base input resistance

From Fig. 1.34c,

$$R_{in} = \frac{v_{in}}{i_{in}} \div \frac{-h_{ie}i_1}{-i_1 - h_{fe}i_1} = \frac{h_{ie}}{1 + h_{fe}}. \tag{1.34}$$

The input resistance for the common base connection is much less than for the common emitter, being 49 Ω if h_{ie} is 2 kΩ, and h_{fe} is 50.

This low resistance is attributable to the increase in input current caused by $h_{fe}i_1$, passing through the input circuit.

Common collector input resistance

Applying Ohm's Law and KCL to Fig. 1.34c,

$$R_{in} = \frac{v_{in}}{i_{in}} \div \frac{h_{ie}i_1 + v_o}{i_1} = \frac{h_{ie}i_1 + (1 + h_{fe})i_1R_L}{i_1},$$

$$= h_{ie} + (1 + h_{fe})R_L. \tag{1.35}$$

If R_L is greater than zero, the input resistance for the common collector stage is higher than for common emitter operation, since the addition of the output voltage to the $h_{ie}i_1$ term increases the input voltage for a fixed input current. However, even though the input resistance can be very high (several MΩ), the base of the transistor must always draw a current which is equal to the output current divided by the current gain. (Reverse biased input junction devices, such as the field-effect transistor, have a high input resistance and do not require any input current.)

For $R_L = 1$ kΩ, $h_{fe} = 50$ and $h_{ie} = 2$ kΩ,

$$R_{in} = 53 \text{ k}\Omega.$$

An approximation frequently used for the emitter follower input resistance is $R_{in} \doteq h_{fe}R_L$.

Input resistance curves are shown in Fig. 1.37.

Output resistance

In common emitter operation the output resistance remains relatively constant as the source resistance R_S changes, but varies for the other two connections as shown in Fig. 1.38.

In Fig. 1.38b, and for common emitter operation, the current generator provides a current which opposes the output current, thus

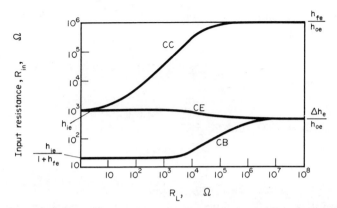

Fig. 1.37. Comparison of input resistance. For common base and common emitter r_{in} tends to the same value, as R_L becomes very large. For common collector, input resistance is proportional to R_L over a wide range of values.

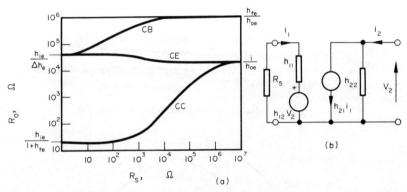

Fig. 1.38. Output resistance as a function of source resistance. $r_{out} = v_2/i_2$, i.e. the resistance seen when looking into the output terminals.

increasing the output resistance when base current flows. The opposite effect occurs with common base and common collector connections, as h_{fb} and h_{fc} are both negative,

$$i_1 = \frac{-h_{12}v_2}{h_{11}+R_S} \quad \text{and} \quad i_2 = i_1 h_{21} + v_2 h_{22}.$$

Thus, output conductance,

$$G_o = \frac{i_2}{v_2} = \frac{-h_{12}h_{21}}{h_{11}+R_S} + h_{22}. \tag{1.36}$$

If the "h" parameters for common collector operation from Table 1.1 are substituted in eqn. (1.36),

$$G_{oc} = \frac{(1 + h_{fe})(1 - h_{re})}{h_{ie} + R_S} + h_{oe},$$

$$\doteqdot \frac{1 + h_{fe}}{R_S} \quad \text{for} \quad \frac{1 + h_{fe}}{h_{oe}} > R_S > h_{ie}.$$

Thus for $R_S = 5 \text{ k}\Omega$, and $h_{fe} = 49$,

$$R_{oc} \doteqdot 100 \ \Omega.$$

Power gain

The output power divided by the input power gives the power gain. Thus, for resistive loads,

$$\text{Power gain} = \frac{v_o i_o}{v_{in} i_{in}} = A_v A_i.$$

The power gain is the product of voltage and current gains.

It is apparent that the common emitter will have the greatest power gain as it has the highest voltage and current gains. The power gains are compared in Fig. 1.39.

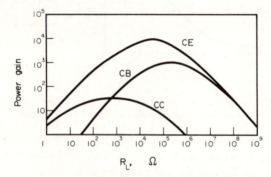

Fig. 1.39. Power gain as a function of load resistance, for the case where the signal source is matched to the input resistance of the transistor.

Performance of transistor amplifiers

Because of its ability to provide both voltage and current amplification the common emitter configuration is the one generally used in transistor amplifiers. It is also preferred because its input and output resistances are normally of a more useful magnitude and are reasonably constant. In tandem connection the common emitter (due to the current gain) can provide voltage amplification. If a is the current gain of each stage, then, referring to Fig. 1.40,

$$v_1 = -a_1 \cdot r_{in2} \cdot i_{in} = \frac{-a_1 \cdot r_{in2}}{r_{in1}} \cdot v_{in}. \qquad (1.37)$$

$$v_L = a_1 a_2 R_L i_{in},$$

$$= \frac{a_1 a_2}{r_{in1}} \cdot R_L v_{in}, \qquad (1.38)$$

if

$$r_{in1} = r_{in2}.$$

For common emitter operation, if the output and load resistances are equal, the voltage gain is equal to the current gain which will generally approach β, being in the range 50 to 250. If the load resistance is very much greater than r_{in}, the gain is increased providing a_2 is not significantly reduced.

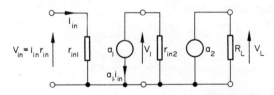

Fig. 1.40. Approximate equivalent network of two transistors in tandem.

For common base, with equal input and load resistances, the current gain is less than unity and there will be no voltage gain. If R_L is greater than r_{in} there will be a voltage gain, but stages in tandem provide no increase in gain unless impedance transformation is provided (transformer connection). For common collector, the input resistance is a function of load resistance, $r_{in} \doteqdot a_1 a_2 R_L$. Thus,

$$v_L \doteqdot \frac{a_1 a_2 R_L}{r_{in1}} \cdot v_{in} \doteqdot \frac{a_1 a_2 R_L}{a_1 a_2 R_L} \cdot v_{in} = v_{in}. \qquad (1.39)$$

1.17.   Examples

(a) *Common emitter*

Figure 1.41a represents a three-stage direct coupled amplifier, each transistor having the h parameters, $h_{21} = \beta = 50$, $h_{11} = 1\,\text{k}\Omega$, $h_{12} = 5 \times 10^{-4}$ and $h_{22} = 50 \times 10^{-6}$. The hybrid representation of a single stage is drawn in Fig. 1.41b.

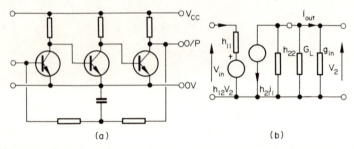

Fig. 1.41. The three-stage d.c. amplifier with an equivalent network of one stage.

Due to the low input resistance v_2 is small, and since h_{12} is also small the voltage generator $h_{12}v_2$ may often be neglected for approximate calculations.

The current gain

$$\frac{i_{\text{out}}}{i_{\text{in}}} = \frac{h_{21}g_{\text{in}}}{h_{22} + G_L + g_{\text{in}}}. \tag{1.40}$$

The output voltage

$$v_2 = \beta i_{\text{in}} \times \text{shunt resistance}$$

$$= \frac{h_{21}}{h_{22} + G_L + g_{\text{in}}} \cdot i_{\text{in}}. \tag{1.41}$$

As g_{in} is very much greater than G_L and h_{22} there is efficient current transfer.

Stage current gain, $a = 40$, so overall current gain $= (40)^3 = 64,000$.

Transfer resistance $= v_{\text{out}}/i_{\text{in}} \doteqdot a^3 \cdot R_L \doteqdot 3.5 \times 10^8 \Omega$.

Voltage gain

$$\frac{Z_t}{r_{in}} \doteq \frac{a^3 R_L}{r_{in}} \doteq \frac{350 \times 10^6}{1000} \doteq 3.5 \times 10^5.$$

(b) Common collector

In the stabilized power supply of Fig. 1.42, V_0 is to be held constant at 3 V for all values of current between 0 and 1 A. The

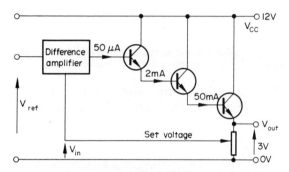

Fig. 1.42. Regulated power supply using a compound common collector stage. The excess voltage $V_{CC} - V_0$ appears across the compound stage.

output transistor must therefore be capable of dissipating a maximum power of $9\,V \times 1\,A = 9\,W$. When drawing 1 A at 3 V the effective load resistance is obviously 3 Ω. This does not appear across the difference amplifier output since the input resistance to the common collector current amplifier,

$$r_{in} \doteq 3 \times h_{f1} \times h_{f2} \times h_{f3}.$$

Given that $h_{f1} = 40$, $h_{f2} = 25$, and $h_{f3} = 20$, $r_{in} = 60\,k\Omega$.
The difference amplifier is therefore only slightly loaded by the current amplifier.

(c) CE–CB–CC d.c. amplifier (Fig. 1.43)

The common collector output stage provides a low impedance voltage output. The input resistance of this stage, for $R_E = 5\,k\Omega$, is $\beta_3 R_E = 200\,k\Omega$. Thus the effective load presented to the common base transistor is 100 kΩ. Since the current gain of such a stage is unity, its transfer resistance R_{T2} is also 100 kΩ. Current gain

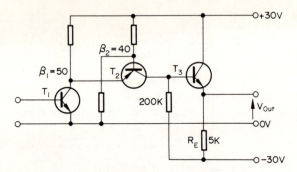

Fig. 1.43. A d.c. amplifier using common emitter, common base and common collector stages in tandem.

provided by stage 1 is approximately equal to β_1 of T_1 since this transistor feeds into the low input resistance of the common base stage. Thus the over-all transfer resistance (output voltage divided by input current)

$$R_T = \beta_1 R_{T2} = 5 \text{ M}\Omega.$$

This implies that 1 μA of input current gives rise to an output of 5 V.

Table 1.2.

	Common emitter	Common base	Common collector
A_v	$\dfrac{-h_{fe}R_L}{h_{ie} + \Delta h_e R_L}$	$\dfrac{(h_{fe} + \Delta h_e)R_L}{h_{ie} + \Delta h_e R_L}$	$\dfrac{(1 + h_{fe})R_L}{h_{ie} + (1 - h_{re} + h_{fe} + \Delta h_e)R_L}$
	$\dfrac{h_{fe}}{1 + h_{oe}R_L}$	$\dfrac{h_{fe}R_L}{h_{ie}}$	$\dfrac{h_{fe}R_L}{h_{ie} + h_{fe}R_L}$
A_i	$\dfrac{h_{fe}}{1 + h_{oe}R_L}$	$\dfrac{-(h_{fe} + \Delta h_e)}{1 - h_{re} + h_{fe} + \Delta h_e + h_{oe}R_L}$	$\dfrac{-(1 + h_{fe})}{1 + h_{oe}R_L}$
	h_{fe}	$\dfrac{-h_{fe}}{h_{fe} + h_{oe}R_L}$	$-h_{fe}$
R_{in}	$\dfrac{h_{ie} + \Delta h_e R_L}{1 + h_{oe}R_L}$	$\dfrac{h_{ie} + \Delta h_e R_L}{1 - h_{re} + h_{fe} + \Delta h_e + h_{oe}R_L}$	$\dfrac{h_{ie} + (1 - h_{re} + h_{fe} + \Delta h_e)R_L}{1 + h_{oe}R_L}$
	h_{ie}	$\dfrac{h_{ie}}{h_{fe}}$	$h_{ie} + h_{fe}R_L$
R_o	$\dfrac{h_{ie} + R_s}{\Delta h_e + h_{oe}R_s}$	$\dfrac{h_{ie} + (1 - h_{re} + h_{fe} + \Delta h_e)R_s}{\Delta h_e + h_{oe}R_s}$	$\dfrac{h_{ie} + R_s}{1 - h_{re} + h_{fe} + \Delta h_e + h_{oe}R_s}$
	$\dfrac{1}{h_{oe}}$	$\dfrac{h_{ie} + h_{fe}R_s}{h_{oe}R_s}$	$\dfrac{h_{ie} + R_s}{h_{fe}}$

1.18.   Summary of the characteristics of transistor amplifiers in terms of h parameters

For the three basic configurations of transistor amplifiers, Table 1.2 sets out input and output resistance, and voltage and current gain, in terms of h parameters. In this table the top expressions in each case are exact, while the lower expressions are approximate.

CHAPTER 2

SCR–UJT–FET

INTRODUCTION

This chapter is concerned with three specific devices, the thyristor or *silicon-controlled rectifier*, the *unijunction transistor* and the *field effect transistor*. The SCR is essentially a semiconductor switch which is used in a wide variety of applications. It enables large currents to be controlled with small input signals. The UJT, on the other hand, with the advent of integrated circuits, is now probably of greater interest to the experimenter than to the professional electronics design engineer. Of the three devices, the FET is the most significant since, apart from its use as a discrete component, the latest form of integrated circuits make use of the same method of construction and employs the same principle of action.

2.1.   The silicon-controlled rectifier[19]

This is a *four-layer* silicon device having the construction shown in Fig. 2.1a. The equivalent structure of Fig. 2.1b shows how it may be represented by a *pnp–npn* complementary pair of transistors

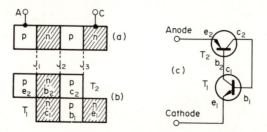

Fig. 2.1. The thyristor is a four-layer device whose action may be simulated by a complementary pair of bipolar transistors.

connected as shown, and having current gains α_1 and α_2 respectively.

The four-layer device has three junctions, J_1, J_2 and J_3. With an applied bias V_{AC}, which makes A (the anode) positive with respect to C (the cathode), junctions J_1 and J_3 are forward biased. J_2 is reverse biased, and the current through it is made up of three components, $I\alpha_1$ the electron current due to T_1, $I\alpha_2$ the hole current due to T_2 and I_{CO} the leakage current,

$$I = I\alpha_1 + I\alpha_2 + I_{CO},$$

from which

$$I = \frac{I_{CO}}{1 - (\alpha_1 + \alpha_2)}. \tag{2.1}$$

If $(\alpha_1 + \alpha_2)$ is very small I very nearly equals the leakage current, and this represents the "off" condition. To turn the device "on", $(\alpha_1 + \alpha_2)$ is increased until the sum equals unity, then I is limited only by external circuit resistance. In the full conduction state the voltage dropped across the SCR is approximately that dropped across a pn junction.

From transistor theory it is known that at low emitter currents α is low, but increases rapidly with increasing I_E. This is shown in Fig. 2.2a while Fig. 2.2b indicates how α varies with an increasing

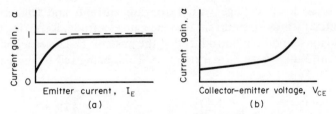

Fig. 2.2. The current gain α, of a transistor, plotted against (a) emitter current and (b) collector–emitter voltage.

collector–emitter voltage V_{CE}. If the anode to cathode voltage across the SCR is progressively increased, eventually junction J_2 will suffer avalanche breakdown. The resulting increase in current causes α_1 and α_2 to increase and the device turns on. The value of V_{AC} which causes this is called the *breakover voltage* V_{BO}. Thereafter the device remains in the on-state provided that the current is sufficient to maintain the sum $(\alpha_1 + \alpha_2)$ equal to unity. This is called the *holding current* I_H. The V/I characteristic for a silicon controlled rectifier is given in Fig. 2.3.

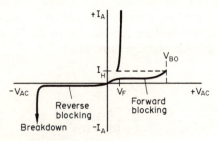

Fig. 2.3. Thyristor characteristic. Note that over a given voltage range the device blocks current in both directions.

Gate control

A third electrode, the gate, forms a connection to the internal p region as shown in Fig. 2.4. Current introduced at this point causes an increase in α_1 and α_2, which is additive to the effect of the anode to cathode voltage V_{AC}. Figure 2.5a illustrates how increasing values of gate current I_G reduces the magnitude of V_{AC} at which breakover occurs. In typical operation the rectifier is biased well below V_{BO}, the breakdown voltage with zero gate current, and triggering is effected by injecting current into the gate electrode. Referring to the analogous circuit, I_G introduced at the base of T_1 causes a collector current $\beta_1 I_G$, where $\beta_1 = \alpha_1/(1 - \alpha_1)$. This, being fed to the base of

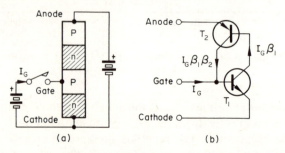

Fig. 2.4. Thyristor gate control. (a) The gate electrode is connected to the internal p region. (b) The two-transistor analogy.

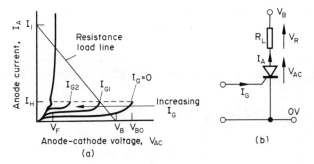

Fig. 2.5. The greater the magnitude of gate current the lower is the anode voltage at which breakover occurs.

T_2, causes a collector current $\beta_1\beta_2 I_G$ which is fed back. The effect is regenerative and the device switches on. Once this has happened the gate ceases to have any control over current flow.

2.2. Switching off

To turn the SCR off the current must be reduced to a value less than I_H. For small values of load current a negative pulse at the gate can be effective since it momentarily reduces α_1. Too large a negative voltage however can damage the device and it is normally restricted to less than 5 V. More typically it is necessary to either open circuit the main current path, or reduce V_{AC} to zero, as illustrated in Fig. 2.6.

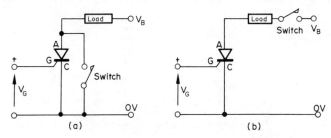

Fig. 2.6. Thyristor switching-off methods. Operation of the switch (a) reduces V_A to zero and (b) open circuits the load current path.

Shunt commutation

The function of the switch in Fig. 2.6a can be performed by an active device; in Fig. 2.7 an *npn* transistor is used. The silicon diode in this circuit prevents the large reverse current which could flow if a gate trigger signal is applied when the SCR is reverse biased. The

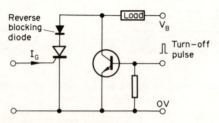

Fig. 2.7. The thyristor can be switched off by the use of a shunt transistor. Note the use of a diode to prevent reverse current flow.

device is switched off when the *npn* transistor is driven into saturation by a positive pulse at its base. An alternative method, shown in Fig. 2.8, makes use of a second controlled rectifier. Each device is switched off by switching the other on, so the two should not be triggered at the same time. Load current is controlled by SCR 1 which must be suitably rated, while since SCR 2 merely performs the switching-off function, it can be a smaller unit. R_2 should be large compared with the load resistance, typically being 10 to 50 times R_L.

With SCR 1 in the on-state, C charges up to $V_B - V_F$, where V_F is the forward voltage drop across the device when conducting. When SCR 2 is switched on, its anode voltage falls to V_F and the charge on C reverse biases SCR 1 causing it to switch off. The $R-L$ network

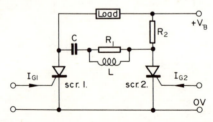

Fig. 2.8. The effect of one thyristor switching-on is to switch the other off. Reverse recovery time is limited by the inclusion of the inductance.

in series with C serves to limit the *reverse recovery current* of the rectifier, as discussed in § 2.3, and has typical values of 100 Ω and 10 μH. The value of C is a function of the load impedance and the recovery time of the device.

Its minimum value is defined by

$$C \geqslant \frac{1.5\, t_{\mathrm{off}}(\mu\mathrm{s})}{Z(\Omega)}\,\mu\mathrm{F}. \qquad (2.2)$$

Series commutation

The essential feature of series commutation is the inclusion of a capacitor C in series with the rectifier, as in Fig. 2.9. When the rectifier is triggered a pulse of current flows to charge the capacitor.

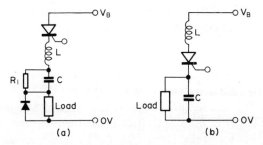

Fig. 2.9. Series commutation methods.

The effect of the inductance L is to cause C to charge to a voltage greater than V_B so that the SCR becomes reverse biased and conduction ceases. Before it can be re-fired C must be discharged and this is done by R_1 in Fig. 2.9a. In Fig. 2.9b the load draws on the energy stored in the capacitor. In both circuits the related time constants must be such that C maintains the reverse bias condition longer than the inherent turn-off time of the device. Series commutation is particularly useful in pulse type applications.

2.3.   Switching characteristics[20]

Transient switching

In any electrical system in which switching occurs, the operation of solenoids, uniselectors, load switching, etc. can give rise to transient voltages in the main supply line and hence to V_B. It has been shown that the depletion layer of a semiconductor junction has associated with it a capacitance, and a transient spike in V_{AC} will cause a current to flow, charging this capacitance. The value of the charging current $i = C\,dV/dt$, can be sufficient to cause α_1 to increase and switch a controlled rectifier from the "off" to the "on" state. The important parameter, in this respect, is the rate of change of off-state voltage, and for a given device the data sheet gives the critical value of dV_{off}/dt as a function of junction temperature. Where it is important to avoid non-gated switching it is good practice to employ some form of transient suppression.

Turn-on time

Ratings of controlled rectifiers are based upon the amount of heat they can safely generate. When an SCR is triggered by a gate pulse, the turn on time t_{gt} is defined as the time interval between the initiation of the gate pulse and the time at which, for a resistive load, current reaches 90 per cent of maximum value. The turn-on time is made up of two parts, a delay time t_d and a rise time t_r, as illustrated in Fig. 2.10a. It is mostly influenced by the value of gate trigger-pulse current; the larger is I_{GT}, the shorter is the turn-on time. During the turn-on period V_{AC} falls from V_{BO} to V_F and current increases rapidly. The resulting high current density can cause localized hot spots within the device. This is particularly significant when currents of large magnitude and narrow pulse width are being switched and it is important that the power dissipated during the turn-on process should be within the limits prescribed for the rectifier.

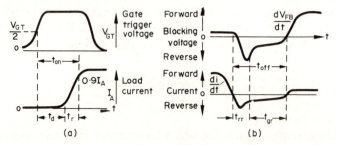

Fig. 2.10. Thyristor switching times. (a) Switching-on, and (b) switching-off.

Turn-off time

This too consists of two parts, as shown in Fig. 2.10b, a *reverse recovery time* t_{rr}, and a *gate recovery time* t_{gr}. When a conducting rectifier is switched off, the application of reverse bias causes a reverse current to flow. The reverse recovery time is that taken for this current to overshoot and reach a steady state. It is then followed by the gate recovery time, a period which must elapse before the depletion layer of junction J_2 is again sufficiently established to block bias voltages of values up to V_{BO}. The total time, from the initiation of reverse recovery current to the start of the forward blocking voltage is called the *circuit-commutated* turn-off time t_q.

The main factors which influence turn-off time are junction temperature, on-state current and the rate of change of current, di/dt, during the forward to reverse transition. However, forward blocking voltage, gate trigger level and the rate of change of the reapplied forward bias voltage, dV_{FB}/dt, also have effect.

Where an SCR is used in a mains frequency (50 or 60 Hz) half wave rectifier circuit, the negative half cycle of the sinewave is more than adequate for complete turn-off. In the case of the full wave bridge rectifier of Fig. 2.11, however, there is no reverse voltage available for turn-off so it is necessary to reduce the bridge output to zero volts, or reduce the load current to less than I_H.

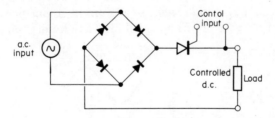

Fig. 2.11. Thyristor controlled bridge rectifier. The action of the rectifying circuit is described in Chapter 7.

2.4. Applications

A basic use of the silicon-controlled rectifier is to control the amount of power transferred from a source to a load.

Control of a.c. power

If for a single rectifier V_{AC} is sinusoidal, it switches off during each negative half cycle provided that the period $1/2f$ is greater than t_q. If the gate current is also an alternating quantity of the same frequency, then reference to Fig. 2.12 shows how the value of I_{GT},

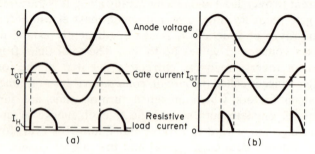

Fig. 2.12. Phase control of the firing angle.

the gate current which causes triggering, determines the proportion of the positive half cycle for which conduction occurs. Obviously this proportion can be varied by adjusting the phase relationship between the gate and bias signals, as shown. However, in general, triggering during negative half cycles of bias signal should be avoided. A way of achieving this is to use a positive pulse at the gate, as is illustrated in Fig. 2.13 for the case where two thyristors are used as an *anti-parallel pair*. Load current commences with the incidence of the gate pulse and is cut off when it falls below I_H. The

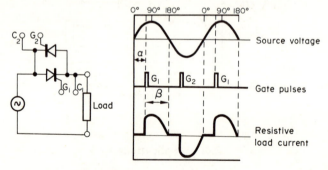

Fig. 2.13. Reverse parallel arrangement for the control of a.c. power. α is the firing angle and β the conduction angle.

delay α of the pulse is termed the firing angle; alternatively the condition can be described in terms of the conduction angle β. By controlling the firing angle it is possible to obtain any value of r.m.s. current from zero to the maximum value available. For a sinusoidal signal controlled in this manner however, the simple relationship between peak and r.m.s. values of current no longer holds (see Fig. 2.14).

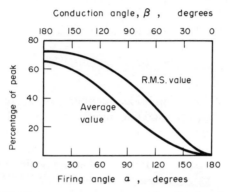

Fig. 2.14. Relationship between r.m.s., average and peak values for a range of firing and conduction angles.

Apart from *phase control*, another method of controlling a.c. power is by means of *integral cycle control*. In this the gate pulses are applied just after the source voltage passes zero, and in a given period (the averaging period) a number of complete cycles are switched to the load. The method is applicable for the supply of power to loads having long time constants, which can "average out" the input power provided.

Control of d.c. power

In Fig. 2.15 the SCR is used as a *chopper*. In the on-state the voltage across the load is $V_B - V_F$, and in the off-state it is zero. For an equal *mark-space* ratio, that is with the on-state and off-state periods equal, the mean voltage across the load is $\frac{1}{2}(V_B - V_F)$. For any other mark-space ratio the mean value is given by

$$V = (V_B - V_F)\left(\frac{t_{\text{on}}}{t_{\text{on}} + t_{\text{off}}}\right). \tag{2.3}$$

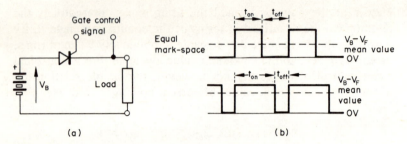

Fig. 2.15. The thyristor as a chopper for d.c. power control.

Controlled rectification

Full wave, half wave and bridge rectifiers, as explained in Chapter 7 on power supplies, are used to convert a.c. to d.c. Figure 2.11 gives one method by which the power transferred from an a.c. source to a d.c. load can be controlled. Alternatively the bridge diodes may be replaced by thyristors as in Fig. 2.16. During positive half cycles of

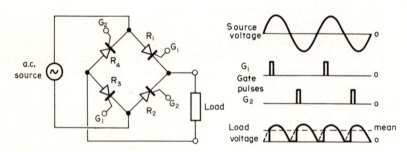

Fig. 2.16. Fully controlled bridge rectifier. Compared with Fig. 2.11, the thyristors carry only one-half the load current.

source voltage, R_1 and R_3 can conduct with the conduction period determined by the phase relationship of gate pulses G_1. During negative half cycles R_2 and R_4 conduct under the control of gate pulses G_2. The conduction angle β determines the amount of d.c. power which is developed in the load.

2.5. Load effects

Resistive load

In Fig. 2.5a a load line has been drawn whose slope, $V_B/I_1 = R_L$, the load resistance. At all times the thyristor bias voltage V_{AC} equals the applied voltage V_B, less V_R the voltage dropped across the load resistance

$$V_{AC} = V_B - V_R.$$

The load line drawn indicates that the *minimum* value of gate current to trigger the device is I_{G1} and, at the point of firing, $V_R = V_B - V_{BO}$, the voltage dropped due to the current I_H flowing through R_L. When in full conduction

$$V_R = V_B - V_F.$$

Capacitive load

If an SCR operates into a large capacitance, as in Fig. 2.17a, when the device fires the capacitor charging current $i = C\, dv/dt$ and, for a fast turn on, this can be large. Accordingly, it is normal practice to

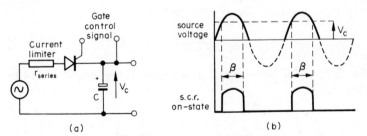

Fig. 2.17. Effect of capacitive load.

include a series resistor, as shown, to limit this current to a magnitude dictated by the device rating. Also, if between conduction periods the capacitor retains its charge V_C, as in the half-wave rectifier, the input voltage must exceed this before conduction can start. The effect is a reduction in conduction angle, as shown in Fig. 2.17b.

Inductive load

The main effect of an inductive load is illustrated for the half-wave rectifier of Fig. 2.18. When the input voltage reverses, the SCR continues to conduct, energy being transferred from the load to the source. After a period θ this current decays and the device switches

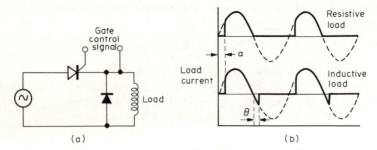

Fig. 2.18. Effect of inductive load. The free-wheeling diode prevents the overshoot and the behaviour is then that of a resistive load.

off. If a *freewheeling diode*[21] is included as shown, when the supply voltage reverses, the load current circulates through this diode, and the rectifier switches off. The output is then the same as if the load were resistive. For an input voltage $E \sin \omega t$ the circuit of Fig. 2.18 provides an average d.c. output voltage,

$$V_L = \frac{E}{\pi \sqrt{2}}(1 + \cos \alpha), \qquad (2.4)$$

while for the full wave rectifier of Fig. 2.19 the output is twice this value.

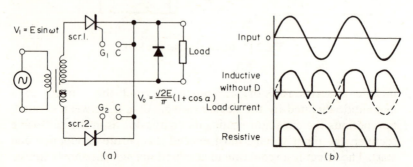

Fig. 2.19. Thyristor control of a full wave rectifier.

2.6. Thyristor ratings

Selection of a thyristor for a given application is made on the basis of its maximum current and voltage ratings, as set out in the manufacturer's data sheets. The current ratings are a function of the working temperature of the device and a maximum permitted temperature is stated.

Mean current rating

The maximum value of current the device can handle is stated as a function of temperature and conduction angle. The angle usually quoted is 180°, the condition obtained when a single-phase resistive load circuit is triggered as soon as the anode swings positive.

Surge current rating

This is intended to indicate the ability of a rectifier to withstand overloads and momentary short circuits under fault conditions. The parameter is given as a peak value of current for one cycle of applied sinusoidal anode voltage at various frequencies, typically 50, 60 and 400 Hz. Additionally surge current curves are provided, showing how these values are derated if the fault condition is maintained over more than one cycle.

Voltage ratings

These are given for the forward and reverse directions and define the continuous peak voltage which the device can handle without breaking down. The values given are for zero gate voltage. If the rating is exceeded in the forward direction the device will break over and will not be damaged. In the reverse direction, however, an excessive voltage will cause avalanche breakdown and the resulting reverse current could destroy the rectifier.

Temperature rating [22]

Heat is generated in a thyristor as a result of power loss during conduction, quiescent leakage current, switching losses during turn-off and turn-on, and gate losses. It is necessary to dissipate this heat in order to maintain the junction temperature below its peak rated value. The relative ease with which heat is conducted from one point to another is given quantitatively in the form of *thermal resistance*. This is the resistance which a device offers to heat flow and is stated in units of degrees centigrade per watt. If T_j is the junction temperature of a thyristor, and T_c is the case temperature, the thermal resistance is

$$\theta_{jc} = \frac{T_j - T_c}{P} \quad °C/W,$$

where P is the power flow in watts from junction to case. Similarly the thermal resistance between the case and ambient, for an ambient temperature of T_a is given by

$$\theta_{ca} = \frac{T_c - T_a}{P} \quad °C/W. \tag{2.5}$$

For a given thyristor the data sheet will state either the maximum permitted junction temperature or the maximum permitted case temperature, and will also include two rating charts of the form given in Fig. 2.20. One indicates the power dissipation within the device for a given mean forward current, and from the other it is possible to determine the maximum permitted case temperature for various conditions of current and conduction angle. With this information it is possible to deduce the thermal resistance of the *heat sink* which is necessary for operation in any given ambient.

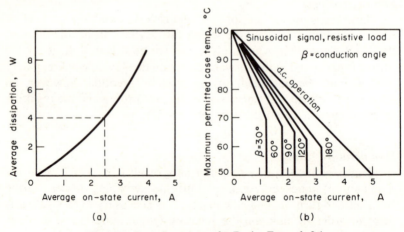

Fig. 2.20. Data sheet curves for Design Example 2.1.

Heat sinks are commercially available with specified values of thermal resistance.

Design Example 2.1

It is required to control the power supplied from a 240 V r.m.s. 50 Hz source to an 80 Ω load resistance in an ambient temperature of 45°C. The conduction angle is to be variable between 30° and 120°.

The arrangement is that of Fig. 2.13, in which each thyristor can conduct during alternate halves of a sine wave.

$$\text{Peak voltage} = 240 \text{ V} \times \sqrt{2} = 340 \text{ V}.$$
$$\text{Peak current} = 340 \text{ V}/80 \text{ } \Omega = 4.25 \text{ A}.$$

Reference is now made to Fig. 2.14. At a conduction angle of 30° the r.m.s. value of current is 12.5 per cent of peak value (0.5 A) and the average value is 5 per cent (0.2 A). Similarly for a conduction angle of 120°, r.m.s. current is 65 per cent of peak current (2.75 A) and average value is 48 per cent (2 A.)

The data sheet for an R.C.A. thyristor type 40554 gives as absolute maximum ratings:

$$\text{Repetitive peak reverse voltage} = 400 \text{ V}.$$
$$\text{On-state r.m.s. current} = 5 \text{ A at } 60° \text{ and } 60 \text{ Hz}.$$

The device is obviously capable of meeting the specification.

Cooling

Referring to Fig. 2.20a, and allowing for 2.5 A average current, the average forward power dissipation is 4 W. Also, referring to Fig. 2.20b for a conduction angle of 120°, the maximum permitted case temperature is 70°C. Thus, assuming a good thermal contact between the device and its heat sink,

$$\theta = \frac{70°C - 45°C}{4 \text{ W}},$$

and the necessary heat sink must have a thermal resistance of 6.25°C/W.

2.7.   Gate characteristic

Having selected a thyristor which satisfies the load requirements, the provision of a trigger circuit is the main design task. This is best done with the aid of gate characteristic curves of the form illustrated in Fig. 2.21. Gate current I_G is plotted against V_{GC}, for a typical low to medium power thyristor, and two curves define the boundaries of

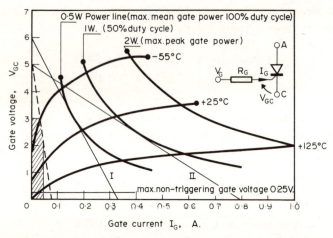

Fig. 2.21. Gate characteristic of a typical low to medium power thyristor.

operating conditions between the temperature limits of $-55°C$ and $+125°C$. Since by Ohm's Law $R = V/I$, any line drawn on a V/I characteristic can, by its slope, represent resistance. Thus the gate input is seen to be resistive, with the $-55°C$ curve indicating maximum gate resistance, and the $+125°C$ curve minimum resistance. Accordingly, the higher the operating temperature, the lower is the gate voltage which is necessary to trigger the SCR.

The shaded area indicates a region in which firing either cannot occur or is uncertain, and any load line should lie to the right of the broken line drawn. The amount of power from a signal source which can be safely dissipated at the gate is dependent upon the proportion of time for which gate current flows. This proportion is stated as a *duty cycle*; if in a repetitive switching operation I_G pulses occupy a quarter of the total time, then the duty cycle is 25 per cent. Three representative power lines have been drawn and, for a given duty cycle, a load line should lie to the left of the relevant power line. Thus, load lines I and II represent the limits for 100 per cent and 50 per cent duty cycles respectively, but as these represent limiting conditions, lines would normally be selected somewhat to the left of them.

Trigger source impedance[23]

The gate drive impedance may be determined using Fig. 2.21. Maximum power is transferred from a source to a load when the impedance of each is equal, and then half the applied voltage is dropped across the load. Let a voltage V_G be applied to the gate through a resistance R_G Ω. Maximum power is transferred when the gate characteristic is such that the voltage dropped is $V_G/2$ and the power transferred is then $(V_G/2)^2/R_G = V_G^2/4R_G$ W. If P_{GM} is the maximum permitted mean gate power then, for a duty cycle of x per cent,

$$\frac{x}{100} = \frac{P_{GM}}{V_G^2/4R_G},$$

from which

$$R_G = \frac{xV_G^2}{400P_{GM}} \ \Omega. \tag{2.6}$$

Example

For $P_{GM} = 0.5$ W, $V_G = 6$ V and a 100 per cent duty cycle,

$$R_G = \frac{100 \times 36}{400 \times 0.5} = 18 \ \Omega,$$

the resistance represented by load line I in Fig. 2.21. Similarly load line II represents the source impedance for the conditions $P_{GM} = 0.5$ W, $V_G = 5$ V and a 50 per cent duty cycle.

Trigger circuit ratings

The characteristic curves, and hence the limiting conditions for operation, vary from unit to unit. There are certain limitations, however, which, in general, apply to all devices and these provide guidelines for design.

The maximum permitted mean power dissipation at the gate is

0.5 W, with a voltage limit of 10 V and a current limit of 2 A. For pulse operation, the allowable peak pulse power is increased as the pulse duration is decreased. Instead of the three duty-cycle power lines of Fig. 2.21 some manufacturers give maximum power lines related to the duration of gate pulses, as, for instance, in Fig. 2.23.

At 25°C no SCR will fire with trigger voltage levels less than 0.25 V, a fact which may be usefully employed. To avoid large reverse currents through the device it is necessary to ensure that triggering does not occur when the anode swings negative. This may be achieved by a diode clamping circuit between the anode and gate,

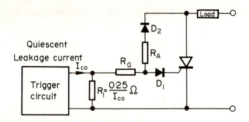

Fig. 2.22. Some protective devices associated with the gate circuit.

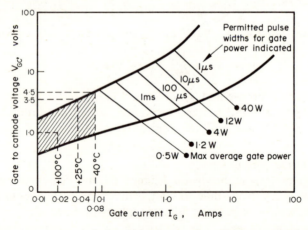

Fig. 2.23. Gate characteristic in which trigger pulse widths are related to lines of maximum gate dissipation.

which limits the gate voltage to 0.25 V when the negative anode swing occurs (D_2 in Fig. 2.22). However, if a gate trigger circuit has a quiescent leakage current then, to avoid misfiring, care must be taken to ensure that this current does not give rise to voltages greater than 0.25 V. A suitably chosen resistor across the trigger circuit output can be used for this purpose (R_1 in Fig. 2.22).

If the gate of an SCR is allowed to go more than 5 V negative, reverse trigger current will flow and the device could be destroyed. In circuits where such a negative voltage swing is possible, it is good practice to include a diode in series with the trigger signal source to prevent reverse trigger current, such as D_1 in Fig. 2.22.

Limitations of trigger circuits

Most thyristor applications are concerned with controlling the amount of power transferred from a source to a load, and many firing circuits are based on the use of a phase-shifted sinewave. Since the firing characteristic varies with temperature the method does not give precise control; where accurate timing is required it is preferable to employ a trigger pulse having a steep-fronted waveform. The pulse duration need only be of the order of 10 μs provided that sufficient voltage is applied to the anode. When controlling a.c. power by varying the conduction angle, however, a very short pulse is unsuitable since, as the conduction angle approaches 180°, V_A becomes very small and the thyristor may fail to fire. A solution is either to use a pulse of longer duration or to limit the conduction angle to about 170°.

Summary

The silicon-controlled rectifier, or thyristor, is a semiconductor switch. It behaves like a normally open mechanical switch for voltages up to a stated value (the breakover voltage). When the switch is closed (by gate control) current flows through it in one direction only. Once open, the switch can only be closed by either reducing the voltage across it to zero or by reducing the current to some value less than I_H, the holding current.

A device which offers considerable advantages in thyristor trigger circuits is the unijunction transistor, and this is now considered.

2.8.   The unijunction transistor[24]

The concept of the unijunction transistor dates back to 1948 and, in those early days, the device was known as a double-base diode. The symbol for the UJT is shown in Fig. 2.24 together with an illustration of the physical construction of the device. It comprises a slice of lightly doped (high resistivity) n-type silicon to which an

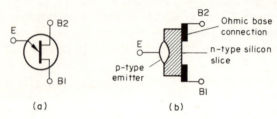

Fig. 2.24. The unijunction transistor. (a) Device symbol, and (b) UJT structure.

emitter connection is made with p-type material. On two sides of this pn junction, ohmic connections are made to provide the two base terminals B_1 and B_2. The ohmic resistance between these, the *interbase resistance* r_{bb}, has typical values between 4.7 kΩ and 9.1 kΩ at 25°C, with a positive temperature coefficient of about 0.8 per cent per degree centigrade. When the value is specified at 25°C the interbase resistance is given the symbol r_{bbo}.

The important feature of the UJT is its negative resistance characteristic. That is, as current through the device increases, voltage across it falls. With suitable circuit arrangement it may therefore be used as an oscillator, as is discussed in § 8.2.

Device action

Consider the simplified model of Fig. 2.25a. Due to the potential divider action of r_{b1} and r_{b2} the diode is reverse biased by a voltage $r_{b1}V_{BB}/(r_{b1}+r_{b2})$. The only current flowing in the input circuit is diode leakage current which, at 25°C, is a few μA. The ratio $r_{b1}/(r_{b1}+r_{b2})$ is termed the *intrinsic stand-off ratio* and is given the symbol η. To cause the diode to conduct, the emitter voltage V_E must be raised to a value $\eta V_{BB} + V_D$ where V_D is the voltage dropped across the diode. At this point the junction becomes

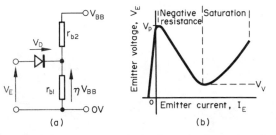

Fig. 2.25. (a) Simple equivalent circuit of a UJT and (b) its characteristic curve.

forward biased and minority carriers are injected into the n-type silicon slice. These are swept into the region of base 1 and cause the diode to become even more forward biased. The emitter current I_E then increases regeneratively until it is limited by the source impedance.

A plot of V_E against I_E is given in Fig. 2.25b and three distinct regions are identified. The first represents the off-state, and shows the leakage current and the peak voltage

$$V_p = \eta V_{BB} + V_D, \qquad (2.7)$$

the point at which conduction starts. Since in the second region current increases as voltage falls, a negative resistance characteristic is indicated, and this extends to some valley voltage V_v. At this point, hole injection from the emitter is so heavy that the effective base 1 resistance r_{b1} has been reduced to a minimum value r_{sat} which is of the order of 20 Ω for an emitter current of 50 mA. A further increase in I_E results in the saturation region of the curve in which the dynamic resistance is again positive. The two conditions of conduction require different equivalent circuits and these are given in Fig. 2.26.

It will be noticed that for any given value of emitter current there is a unique value of emitter voltage. However, only for emitter voltages less than the valley voltage is I_E single-valued, and this occurs in the off-region of the V_E/I_E curve. For values of V_E greater than V_v, I_E is multivalued. Thus, if when the device is conducting, the emitter voltage is reduced to a value less than V_v, the device will switch off.

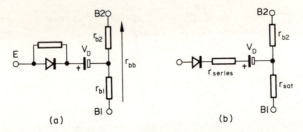

Fig. 2.26. Unijunction equivalent circuits for (a) the negative resistance region and (b) the saturation region of the characteristic curve.

Applications

The UJT is characterized by a stable triggering voltage (V_p), a low value of firing current (I_p), a stable negative resistance characteristic and a high pulse current capability. It is therefore suitable for inclusion in oscillators, timing circuits, voltage and current sensing circuits and thyristor triggering circuits.

Load lines

In Fig. 2.27 two load lines have been drawn. Load line I intersects the curve only at *B* in the negative resistance region. Load line II has three points of intersection, at *A* in the off region, at *B*, and at *C*

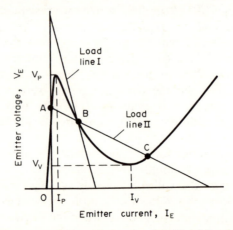

Fig. 2.27. Load lines on the characteristic curve. Load line I is suitable for use in an astable circuit and load line II affords bistable action.

in the saturation region. Points A and C represent inherently stable conditions, while point B is only conditionally stable. If a capacitor is added between the emitter and base 1 the point becomes unstable. In Chapter 9 on Waveform Generators three types of multivibrator are considered. The *astable* type requires no stable states, the *monostable* requires only one, while the *bistable*, as its name suggests, has two stable states. Thus, the two loads represented by lines I and II are suitable for astable and bistable circuits respectively, with the proviso that, for the latter, the capacitor in the astable circuit of Fig. 2.28a is replaced by a resistor as shown. Since the basic configuration of many UJT circuits is that of the relaxation oscillator, the astable circuit will now be considered in detail.

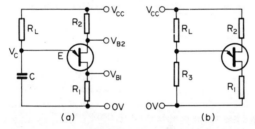

Fig. 2.28. If C is replaced by R_3 the astable arrangement of (a) becomes the bistable arrangement of (b).

Circuit action

The waveforms obtained are shown in Fig. 2.29a, and in Fig. 2.29b the ideal and actual paths of operation are illustrated. The cycle starts at A with capacitor C charging through R_L. When the charge reaches V_p conduction starts and, since the capacitor voltage cannot change instantaneously, V_E is unchanged and operation jumps to point D in the saturation region. C now discharges on a time constant $C(R_L + r_{sat} + r_{series})$, as in Fig. 2.26b, and operation follows the V_E/I_E curve. When the operating point enters the negative resistance region, a continuing fall in I_E requires an increase in V_E which is not available, and the device switches off. The capacitor now starts to recharge and the cycle is repeated. In practice it takes a finite time for I_E to increase to the value at point D and, in that time, the capacitor starts to discharge. This accounts for the difference between the ideal and actual paths of operation. For small

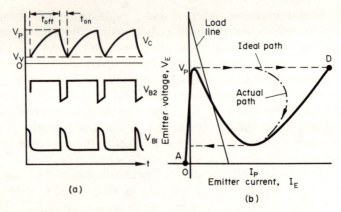

(a)   (b)

Fig. 2.29. Waveforms and operation of a UJT astable multivibrator.

values of R_1 and R_2 the period of oscillation is

$$\tau = 1/f \doteqdot CR_L \cdot \log_e \left(\frac{1}{1-\eta}\right) \text{ s.} \qquad (2.8)$$

Manufacturers' data sheets give values of η in the range 0.47 to 0.61 which for the log term yields 0.64 to 0.94. Thus, as an approximation,

$$\tau = 1/f \doteqdot 0.8CR_L \text{ s.} \qquad (2.9)$$

Design considerations

The value of V_p determines the point at which conduction starts and is the most important characteristic of the device. Since η is a ratio it is unaffected by temperature change. Hence, reference to eqn. (2.7) shows that any variation of V_p due to temperature change is caused by a changing V_D. The effect may be compensated by suitable choice of R_2. The positive temperature coefficient of r_{bb} causes interbase current to decrease with rising temperature. The voltage drop across R_2 therefore falls so that V_{BB} rises and, at some specific temperature, exactly offsets the accompanying fall in diode voltage. The temperature coefficient of r_{bb} is not linear, however, so compensation is not complete for all temperatures. Providing that the value of R_1 is less than 100 Ω, if R_2 is selected using the equation,

$$R_2 = \frac{0.4r_{bbo}}{\eta V_{cc}} + \frac{R_1(1-\eta)}{\eta} \ \Omega, \qquad (2.10)$$

then V_p will equal ηV_{cc} with reasonable accuracy over a temperature range 0°C to 100°C.

Since the frequency of oscillation is a function of the time constant CR_L, the higher the value of R_L, the smaller will be the capacitor, for a given time constant. Normally, the resistor is made as large as possible, consistent with the conditions for oscillation, that is having a load line whose intercepts with the characteristic curve indicates no stable states. As an alternative, the circuit will oscillate if R_L has a value such that

$$\frac{V_{BB} - V_v}{I_v} < R_L < \frac{V_{BB} - V_p}{I_p}. \tag{2.11}$$

DESIGN EXAMPLE 2.2

Required, a UJT astable multivibrator, having a frequency of 100 Hz, operating from a 20 V d.c. power supply.

The device selected is the Texas Inst. 2N 1761, having the characteristic curves shown in Fig. 2.30 for operation at 25°C. The manufacturer's data sheets state minimum and maximum values for most of the necessary parameters. Let the values used be:

$$\eta = 0.5, \quad r_{bbo} = 7 \text{ k}\Omega, \quad I_p = 20 \, \mu\text{A} \quad \text{and} \quad I_v = 8 \text{ mA}.$$

Referring to Fig. 2.28a, R_1 must be less than 100 Ω; let it be 22 Ω.

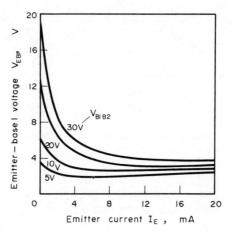

Fig. 2.30. Characteristic curves for Design Example 2.2.

Then, from eqn. (2.10)

$$R_2 \doteq \frac{0.4 \times 7 \text{ k}\Omega}{0.5 \times 20 \text{ V}} + \frac{22 \, \Omega(1 - 0.5)}{0.5} \doteq 300 \, \Omega,$$

and, $V_p = \eta V_{cc} = 10 \text{ V}$.
Also, from Fig. 2.30, $V_v \doteq 3.5 \text{ V}$.
   Thus, using eqn. (2.11),

$$\frac{20 - 3.5 \text{ V}}{8 \text{ mA}} < R_L < \frac{20 - 10 \text{ V}}{20 \, \mu\text{A}},$$

for the circuit to oscillate, and R_L must be in the range 2 kΩ to 500 kΩ. Let it be 100 kΩ. Then, from eqn. (2.9),

$$C = \frac{1}{100 \text{ Hz} \times 0.8 \times 100 \text{ k}\Omega} = 0.125 \, \mu\text{F}.$$

In this example a typical value of $\eta = 0.5$ has been used, since the data sheet indicates a possible production spread from 0.47 to 0.61. To take account of this spread it is necessary to make provision for frequency adjustment. This has been done in Fig. 2.31a, in which R_L has been replaced by a fixed resistor of 75 kΩ in series with a 50 kΩ potentiometer.

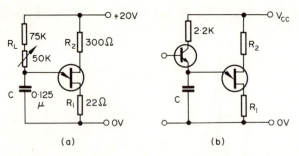

(a)   (b)

Fig. 2.31. (a) The astable multivibrator of Design Example 2.2. (b) If the potentiometer is replaced by a transistor the frequency can be adjusted by varying the base voltage.

Automatic frequency adjustment
   The value of R_L determines the rate at which the capacitor charges, and hence the frequency of oscillation. In Fig. 2.31b R_L has been replaced by an *npn* transistor in series with a resistor, and the

collector current, which is the charging current for the capacitor, is controlled by the base voltage. Thus, if some *error signal* is available the frequency can be adjusted automatically.

2.9.   The UJT for thyristor triggering[25]

The high pulse current capacity of the device makes the UJT astable circuit suitable for thyristor triggering. In this respect, for the control of power from an a.c. source, the astable must be synchronized with the source frequency.

Synchronization

The circuit diagram of Fig. 2.32 is essentially a modified form of the basic bridge rectifier of Fig. 2.11. A UJT astable circuit is used to provide the trigger pulse for the SCR. The effect of the zener diode is to limit the waveform at A so that, at B, the waveform of the

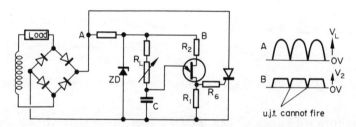

Fig. 2.32. Control of full wave bridge rectifier. The astable frequency is synchronized with the frequency of the power source.

voltage applied across the UJT has the square shape shown. If the zener diode has a breakdown voltage V_z then, for small values of R_1 and R_2, the UJT will fire when $V_p \doteqdot \eta V_z$. The firing point is thus independent of the peak value of bridge voltage.

When the voltage at B falls to zero, the UJT cannot fire. Starting at that point, C starts to charge and, when its voltage reaches ηV_z the device fires, discharging C. The width of the pulse which results across R_1 is the on-time for the device, the time taken to discharge the capacitor through the UJT. This pulse is used to trigger the thyristor, which conducts for the rest of the half cycle, causing power to be transferred from the source to the load. The voltage at

A is now held at V_F, the thyristor forward voltage drop, and the capacitor cannot recharge for the rest of the half cycle. When the half cycle ends the thyristor switches off, the voltage at A and B rises with the beginning of the next half cycle, and the process is repeated. The firing angle is determined by the setting of the resistor R_L, and the value of R_G is computed using eqn. (2.6).

Increased pulse width

A graph given in the data sheets for the UJT relates *emitter fall time* to capacitance. If a greater trigger pulse width is required it is necessary to increase the discharge time of the capacitor. This is done in the circuit of Fig. 2.33. While the capacitor is charging, diode

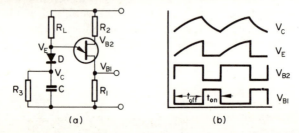

(a)   (b)

Fig. 2.33. The addition of the diode and R_3 enables the pulse width to be increased.

D is forward biased. When the UJT fires, its emitter voltage falls and D becomes reverse biased. C now starts to discharge through R_3, from V_p towards zero. A point will be reached when the capacitor voltage has fallen below the emitter voltage, the diode will again conduct, and the device turns off as before. For this circuit, the conditions for oscillation are:

$$\frac{R_3 V_{BB}}{R_3 + R_L} > V_p, \quad \text{and} \quad R_L > \frac{V_{BB}}{I_v}. \tag{2.12}$$

The period of oscillation is the sum of the on-time and the off-time. If V_1 is defined as the voltage at which the R_L load line crosses the characteristic curve, as in Fig. 2.30, these times are given by:

$$t_{on} = CR_3 \log_e (V_P/V_1) = CR_3 \log_e (V_p/V_v) \text{ s,} \tag{2.13}$$

since in practice there is little difference between V_1 and V_v. Also,

$$t_{off} = \frac{CR_3R_L}{R_3 + R_L} \log_e \left[\frac{1}{1 - \eta\left(\dfrac{R_3 + R_L}{R_3}\right)} \right] \text{s}. \qquad (2.14)$$

The two equations are interdependent, so for design purposes it is convenient to simplify them.

Let

$$\gamma = \frac{R_3}{R_3 + R_L}, \quad \text{and} \quad CR_L = \tau. \qquad (2.15)$$

Then,

$$t_{off} = \gamma\tau \log_e \left(\frac{1}{1 - \eta/\gamma}\right) \text{s} \qquad (2.16)$$

and

$$t_{on} = \frac{\gamma\tau}{1 - \gamma} \log_e (V_p/V_v) \text{s}. \qquad (2.17)$$

For given values of η, V_p and V_v the solutions may be given in normalized form, with t_{off}/τ and t_{on}/τ plotted against γ. This has been done for $V_p = 6$ V, $V_v = 2.5$ V and $\eta = 0.5$, in Fig. 2.34a. From this figure a second graph, Fig. 2.34b, has been constructed relating the ratio t_{off}/t_{on} to γ. The two graphs may be used for design purposes.

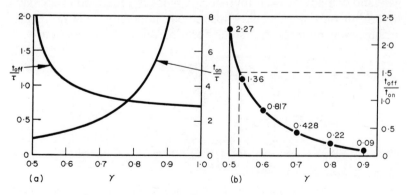

Fig. 2.34. Design graphs for use in Design Example 2.3.

DESIGN EXAMPLE 2.3

Required, a UJT astable multivibrator, having a frequency of 100 Hz, and providing a pulse of 4 ms duration for triggering the anti-parallel pair arrangement of Fig. 2.13.

If a 12 V zener diode is used, as in Fig. 2.32, $V_{CC} = 12$ V $\doteqdot V_{BB}$. Thus, if $\eta = 0.5$, $V_p = \eta V_{CC} = 6$ V. Also, from the data sheets, $I_v = 7.5$ mA, $V_v = 2.5$ V and $r_{bbo} = 7$ kΩ.

From eqn. (2.10), $R_2 = 470$ Ω.

Timing components

The specified frequency is 100 Hz, i.e. a period of 10 ms. Thus, $t_{off} = 6$ ms, $t_{on} = 4$ ms, and the ratio $t_{off}/t_{on} = 1.5$. Reading from Fig. 2.34b,

$$\gamma = \frac{R_3}{R_L + R_3} = 0.53, \quad \text{so } R_3 \doteqdot 1.125 R_L.$$

As a condition of oscillation eqn. (2.12) gives R_L a minimum value of 12 V/7.5 mA = 1.6 kΩ. Also, $R_3 V_{BB}/(R_3 + R_L) > V_p$ and if R_L is selected as 10 kΩ this gives a minimum value for R_3 as 10 kΩ also. Thus $R_3 = 1.125$, $R_L = 11.25$ kΩ satisfies this second requirement.

From eqn. (2.13), 4 ms = $C \times 11.25$ kΩ $\times \log_e (6/2.5)$ from which $C = 0.4$ μF.

Gate connections

The gate signals derived from base 1, must be applied between gate and cathode of each thyristor, A pulse is generated every half cycle, but only the device whose anode is positive will fire. This arrangement requires that some form of thyristor reverse current

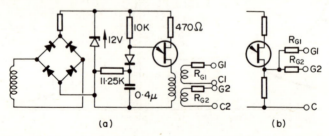

(a)   (b)

Fig. 2.35. (a) Gate drive circuit for thyristor anti-parallel pair. (b) Modified output circuit for the full wave rectifier of Fig. 2.19.

protection is used. For the full-wave rectifier of Fig. 2.19, the thyristor cathodes are commoned so that base 1 circuit of Fig. 2.35b is suitable. For the anti-parallel pair, however, a common cathode line cannot be used, since this would short out both thyristors, hence the gate source for each SCR must be independent. This is achieved by the use of the pulse transformer in the base 1 circuit of Fig. 2.35a. Note that in this example the duty cycle is 40 per cent, and this figure would be used in eqn. (2.6) to compute the values of R_G.

2.10. A bipolar transistor analogy

It is possible to arrange a complementary pair of transistors, as shown in Fig. 2.36a, in such a way that they function in the same manner as a UJT (see also Fig. 2.1). Although several components are used to replace one unijunction transistor the arrangement has

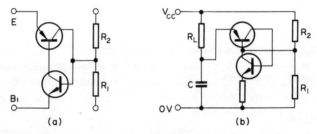

Fig. 2.36. (a) Complementary pair of transistors connected to function as a UJT and (b) as used in an astable multivibrator.

the merit that the intrinsic stand-off ratio can be defined by the choice of R_1 and R_2. The circuit of Fig. 2.36b shows how the complementary pair may be used in an astable multivibrator circuit.

A bipolar device is so called because it makes use of both negative and positive charge carriers. The UJT makes use of only holes and so is a unipolar device. Another unipolar device of particular importance is the field effect transistor.

2.11.   Field effect transistors[26]

The addition of specific impurities to a piece of pure semiconductor material gives rise to a surplus of free charge carriers, and hence lowers its resistivity. The greater the doping level, in n-type material, for instance, the greater the number of free electrons and the lower the resistance of a given sample. Thus n^+ material is more heavily doped than n material and has lower resistivity. On the other hand, if the number of free electrons is *depleted*, resistivity is increased. The action of field-effect transistors makes use of this fact and, in their simplest form, such devices can be looked upon as variable resistors.

There are two types of FET to be considered, the junction field-effect transistor (JFET) and the insulated gate field-effect transistor (IGFET). The latter type is sometimes called a metal–oxide–semiconductor (MOS) transistor. Both devices operate on the same principle, in which current is controlled by an electric field, which changes the resistance between input and output terminals. The two types have different characteristics, however, and must be studied separately.

Junction FET

For an *n-channel* device the starting-point is a bar of n-type silicon with two p-type regions diffused into opposite sides, as shown in Fig. 2.37a. This creates an n-type channel from the input terminal, the *source*, to the output terminal, the *drain*, which passes

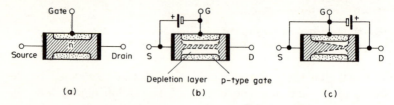

Fig. 2.37. Schematic representation of an n-channel junction FET. The applied bias increases the width of the depletion region and increases channel resistivity.

between the two p regions, the *gate*. If the gate regions are made negative with respect to the n-type substrate the junctions are reverse biased and, as with any pn junction, depletion layers form. Increasing this negative bias causes the depletion layers to spread into the channel where they eventually meet, causing an almost infinite resistance between source and drain. In Fig. 2.37c the gate has been connected to the source and the drain has been made positive with respect to it. Again the junctions are reverse biased but, due to the voltage gradient between drain and source, and the resistive nature of the substrate, one side is biased more than the other and the depletion layer on that side is greater. Initially, increasing the drain voltage, V_{DS}, causes the drain current, I_D, to increase and, with it, the potential gradient. As a result the depletion layer spreads, increasing the drain–source resistance and tending to limit the current. Eventually a point is reached where an increase in V_{DS} causes no further increase in I_D. The drain–source voltage which causes this current limiting condition is known as the *pinch-off* voltage, V_P.

The I_D/V_{DS} characteristic is given in Fig. 2.38, with the gate–source voltage, V_{GS}, as parameter. An increase in gate bias causes

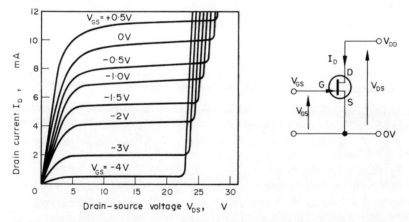

Fig. 2.38. Characteristic curve of an *n*-channel JFET. Note that the gate potential is effective even when it is slightly positive.

pinch-off to occur at a lower value of I_D, because the two effects causing depletion layer spread are additive. In the saturation region, for each curve, I_D remains constant with increasing V_{DS} until avalanche breakdown occurs and I_D increases dramatically. Examination of Fig. D.6 shows the similarity of the characteristic of the FET with that of the thermionic pentode valve. Comparison indicates that the gate is analogous to the control grid, while the source and drain appear to perform the same functions as the cathode and anode respectively. An additional similarity between the two devices is that both have high input impedances compared with a junction transistor which is a low input impedance device. Accordingly there is a similarity with design procedures.

Figure 2.37 is a schematic illustration of the JFET. In practice the device is made using epitaxial planar techniques which imply diffusion from one side only. The construction is therefore as shown in Fig. 2.39. The starting-point is a wafer of relatively thick p-type

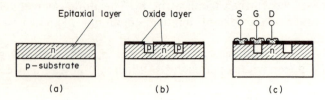

Fig. 2.39. Three stages in the epitaxial planar construction of an n-channel JFET.

substrate (having low resistivity) upon which has been grown a thin epitaxial layer of n-type silicon (a). A high-temperature treatment covers the top surface with silicon dioxide, photo-resist techniques are employed to remove this oxide from the area in which diffusion is to take place, and a p-type diffusion is performed to form the gate (b). A second oxidization and etching process follows and the necessary terminals are made (c). The three drawings of Fig. 2.39 represent a cross-section of the device. Figure 2.40 shows the overall geometry with the oxide layers and terminals removed.

Note that if all the n- and p-type regions are interchanged a p-channel device is obtained.

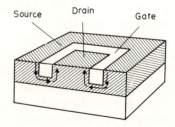

Fig. 2.40. The JFET with oxide layer and contacts removed.

Insulated gate FET (enhancement type)

For this the starting-point is a high resistivity *p*-type substrate into which two low resistivity *n* regions are diffused to form the source and the drain. This is now coated with an oxide layer, then photo-resist and etching operations follow, which allow metallic contacts to be made for the source and drain terminals. At the same time as these contacts are made, a layer of metal is also deposited on top of the oxide layer to cover the channel area. This metallic layer forms the gate which is insulated from the channel by the silicon dioxide. The final arrangement is shown in Fig. 2.41a. Note that in

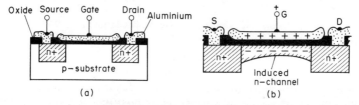

Fig. 2.41. Enhancement mode insulated gate FET.

the absence of any gate voltage the resistance between source and drain is that of two *pn* junctions back-to-back. Note also that a capacitor has been formed between the metal gate and the silicon channel, with the oxide as dielectric.

The operation of the device is illustrated in Fig. 2.41b, and is as follows. If a positive charge is applied to the gate, a corresponding negative charge is induced in the semiconductor material which forms the other side of the gate capacitor. As the positive gate voltage is increased, so is the induced negative charge until the substrate beneath the gate becomes effectively *n* type. Current can flow from source to drain through this "induced" *n*-type channel. The greater the gate voltage is made, the deeper will be the induced channel and the greater will be the source–drain current. In other words, the drain current is *enhanced* by the gate potential and the device is said to be operating in an *enhancement mode*. As before, a reversal of semiconductor polarity will produce a *p* channel, IGFET.

Depletion type IGFET

The structure illustrated in Fig. 2.41a is modified by the inclusion of an *n*-type channel diffused between the source and drain, as in Fig. 2.42a. The application of a negative potential to the gate induces a positive charge in this channel and a depletion layer is formed, increasing the channel resistivity. This is illustrated in Fig. 2.42b. In

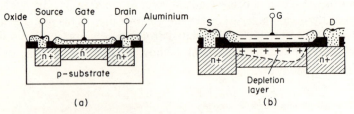

Fig. 2.42. Depletion mode insulated gate FET.

this respect, the characteristics of the device are similar to those of a JFET; the greater the applied negative gate potential, the smaller will be the flow of drain current I_D. The devices differ, however, in their behaviour when the gate is made positive. For the *n*-channel JFET a positive gate bias forward biases the junction and eliminates the depletion layer, so that the gate has no control over the flow of drain current. If, however, the gate of an *n*-channel depletion-type IGFET is made positive, the resulting induced negative charge in the channel has the effect of reducing channel resistivity, and the device can then operate in an enhancement mode (see Fig. 2.43).

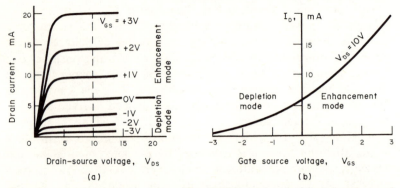

Fig. 2.43. (a) Characteristic curves of an IGFET used in both enhancement and depletion modes of operation. The I_D/V_{GS} curve (b) is obtained by reading off current values for $V_{DS} = 10\,\text{V}$.

Summary of FET types and symbols

Junction field-effect transistors are depletion mode devices, and are therefore in an on-state when no gate potential is applied. The application of a positive potential to the gate of a p-channel device, or a negative potential to the gate of an n-channel device, causes a reduction in drain current. If the potential is progressively increased the device will eventually be turned off.

There are essentially four different types of insulated gate devices. Both the p- and n-channel depletion types are normally in an on-state and require the application of a gate potential to turn them off. On the other hand, p- and n-channel enhancement devices are

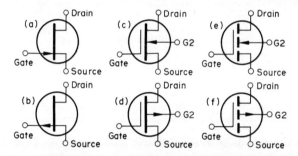

Fig. 2.44. FET symbols. (a) and (b) are JFET, (c) and (d) are depletion mode IGFET and (e) and (f) enhancement mode IGFET. The arrow head points inward for an n-channel device and outward for a p-channel device.

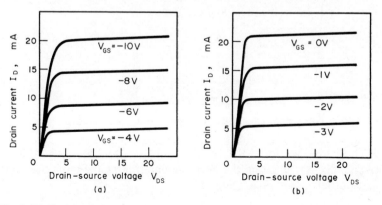

Fig. 2.45. Characteristic curves for (a) a p-channel enhancement IGFET and (b) an n-channel depletion IGFET. Note that the first is normally *off* and the second normally *on*.

normally in the off-state. A p-channel enhancement FET requires a negative gate potential to turn it on, while to turn an n-channel enhancement type on, a positive gate potential is necessary.

The symbols used to represent the six different types of FET are given in Fig. 2.44. In four of these, a fourth terminal G_2 is shown. This makes a connection to the substrate, and is brought out of the device by some manufacturers. The majority of devices available, however, have the substrate internally connected to the source. Of the four different types only two insulated gate devices have been widely adopted; these are the p-channel enhancement mode and the n-channel depletion mode. Typical characteristic curves for these are given in Fig. 2.45.

2.12. FET amplifier characteristics[27]

As the field effect transistor is usually operated with the input junction in the reverse biased mode, input current is very small. For the depletion FET, which is the most commonly encountered type (Fig. 2.38), the characteristics are similar to those of the thermionic vacuum pentode discussed in Appendix D.

Both the junction and insulated gate forms of FET have extremely high input resistance and the h parameter network of Fig. 1.25, which represents a low-input current-operated device, is not a convenient model. Instead, the y parameter form of Fig. 2.46 is used. In this, the dependent energy sources are operated by voltages, and the network is symmetrical.

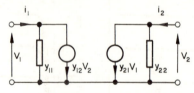

Fig. 2.46. Admittance or y parameter two port network.

The equations for the network can be written down using KCL (Appendix A), equating the currents entering and leaving the input and output nodes respectively. Thus,

$$i_1 = y_{11}v_1 + y_{12}v_2, \tag{2.18}$$

$$i_2 = y_{21}v_1 + y_{22}v_2. \tag{2.19}$$

The parameters are short-circuit admittance elements, as they are evaluated by making either v_1 or v_2 zero, i.e.

$$y_{11} = i_1/v_1 \quad (v_2 \text{ zero}),$$
$$y_{12} = i_1/v_2 \quad (v_1 \text{ zero}),$$
$$y_{21} = i_2/v_1 \quad (v_2 \text{ zero}), \qquad (2.20)$$
$$y_{22} = i_2/v_2 \quad (v_1 \text{ zero}).$$

The high input resistance of the FET means that y_{11} is very small and can usually be neglected. The feedback element, y_{12}, has little effect at low frequency and can likewise be omitted. Thus, in Figs. 2.47, 2.48 and 2.49, the equivalent networks for common source, common gate and common drain modes of operation are shown with these input elements removed.

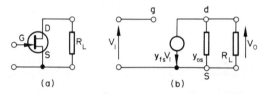

Fig. 2.47 (a) Small signal network for common source operation and (b) equivalent network.

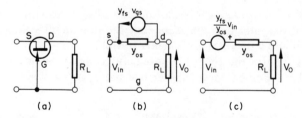

Fig. 2.48. Common gate operation. (a) Small signal network and (b) and (c) equivalent networks.

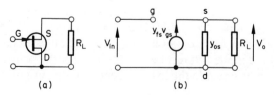

Fig. 2.49. (a) Small signal network for common drain operation and (b) equivalent network.

Common source operation

This is illustrated in Fig. 2.47 and is the basic connection, equivalent to the common emitter operation of a bipolar transistor.

$$y_{21} = y_{fs}, \quad \text{the forward transfer parameter,}$$

and

$$y_{22} = y_{os}, \quad \text{the short-circuit output conductance.}$$

Typical figures for the 2N 3819 are:

$$y_{fs} = 5 \text{ mS}; \quad y_{os} = 50 \text{ } \mu\text{S}.$$

Voltage gain. Figure 2.47b shows that the common source output voltage is produced by the current $y_{fs}v_1$ flowing through y_{os} and the load R_L in parallel, the negative sign resulting from the direction of flow of the voltage-dependent current source.
Thus,

$$v_0 = \frac{-y_{fs}v_1}{y_{os} + 1/R_L}, \tag{2.21}$$

and the common source voltage gain,

$$\frac{v_0}{v_1} = \frac{-y_{fs}R_L}{1 + y_{os}R_L}. \tag{2.22}$$

The curve for voltage gain as a function of load resistance is shown in Fig. 2.50a, and it is apparent that the maximum voltage gain is y_{fs}/y_{os}, if the load resistance tends to infinity.

This is 100 for the 2N 3819 quoted above, and is an order less than would be expected for a bipolar transistor, as shown in Fig. 1.36.

Common gate operation (Fig. 2.48)

The common gate y-parameter equivalent network can be changed to a series arrangement by converting the current source, $y_{fs}g_s$, to a voltage source, $y_{fs}v_{in}/y_{os}$. This is the open-circuit voltage which the current source would develop across y_{os}. The gate-source voltage, v_{gs}, for the common gate connection is seen to be the inverse of the input voltage, i.e. $v_{gs} = -v_{in}$. For the series circuit, the input current is equal to the load current, and the output voltage is

$$v_o = iR_L, \tag{2.23}$$

where, using Ohm's Law,

$$i = \frac{v_{in} + (y_{fs}/y_{os})v_{in}}{(1/y_{os}) + R_L}. \tag{2.24}$$

Voltage gain. Common gate voltage gain,

$$\frac{v_o}{v_{in}} = \frac{(y_{fs} + y_{os})R_L}{1 + y_{os}R_L}, \tag{2.25}$$

which, because of the relative smallness of the y_{os} term, has the same form as eqn. (2.22) for common source voltage gain, except that in this case there is no signal inversion. Figure 2.50 shows the common source and common gate voltage-gain curves as being identical.

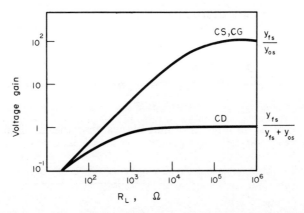

Fig. 2.50. Voltage gain of FET amplifiers as a function of load resistance. For common source and common gate it is essentially the same, except for the signal inversion of the former.

Common drain operation

From the network of Fig. 2.49 the gate–source voltage,

$$v_{gs} = v_{in} - v_o, \tag{2.26}$$

the output voltage,

$$v_o = \frac{y_{fs}v_{gs}}{(1/R_L) + y_{os}}, \tag{2.27}$$

and substituting for v_{gs} from eqn. (2.26),

$$v_o = \frac{y_{fs}R_L v_{in}}{1 + (y_{fs} + y_{os})R_L}. \tag{2.28}$$

Voltage gain. Common drain voltage gain,

$$\frac{v_o}{v_{in}} = \frac{y_{fs}R_L}{1 + (y_{fs} + y_{os})R_L}. \tag{2.29}$$

For R_L somewhat greater than $1/y_{fs}$, the gain is close to unity, which is also the case for the emitter follower or common collector bipolar transistor amplifier. Because of its high input resistance and unity voltage gain, the common drain amplifier or *source follower* is very useful as an isolating or buffer amplifier.

FET amplifier input resistance
This is ideally infinite for the common source and common drain connections, but low for common gate.

From eqn. (2.24), the input resistance for the latter is

$$\frac{v_{in}}{i} = \frac{1 + y_{os}R_L}{y_{fs} + y_{os}}. \tag{2.30}$$

This is plotted in Fig. 2.51 and shows that the minimum input resistance is 200 Ω.

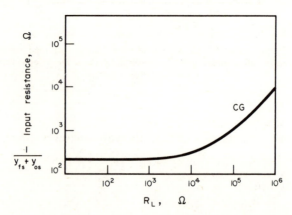

Fig. 2.51. Only the common gate connection has a finite input resistance.

FET amplifier output resistance

Unlike the bipolar transistor amplifier connections, these are independent of the source impedance. From the three voltage-gain equations, the denominator terms show that the output resistance is $1/y_{os}$ for common source and common gate, but is $1/(y_{fs} + y_{os})$ for the common drain connection.

Current gain

As no input current is drawn in the common source and common drain connections, these provide infinite current gain. In the common gate connection, however, the same current passes through the input and the output so in this case current gain is unity.

Table 2.1 summarizes the performance of the basic FET configurations.

Table 2.1. Performance of field-effect transistor configurations

	Common source	Common gate	Common drain
Voltage gain	$\dfrac{-y_{fs}R_L}{1 + y_{os}R_L}$	$\dfrac{(y_{fs} + y_{os})R_L}{1 + y_{os}R_L}$	$\dfrac{y_{fs}R_L}{1 + (y_{fs} + y_{os})R_L}$
Current gain	Infinite	Unity	Infinite
Input resistance	Infinite	$\dfrac{1}{y_{fs} + y_{os}}$	Infinite
Output resistance	$\dfrac{1}{y_{os}}$	$\dfrac{1}{y_{os}}$	$\dfrac{1}{y_{fs} + y_{os}}$

CHAPTER 3

Integrated Circuits

INTRODUCTION

An integrated circuit is comprised of a single silicon chip containing the necessary transistors, diodes, resistors and capacitors, suitably connected to form a complete circuit. The first successful attempt to produce an integrated circuit, in 1959, made use of *mesa* construction, but this method was quickly replaced by the use of planar techniques. Some of the various processes involved have already been mentioned but for ease of reference they are repeated here in greater detail, in a description of the manufacture of an *npn* transistor.

3.1. Manufacturing processes[28]

Epitaxial growth

In order to obtain an acceptable mechanical strength, it was found that the collector region of a mesa transistor (Fig. 1.21) needed to be at least 0.05 mm thick. Additionally, to obtain a low collector capacitance and a high breakdown voltage, it was necessary for the region to have high resistivity. The combination of these two requirements produced an undesirable resistance between the collector junction and the collector contact. The solution to this problem was found in the epitaxial process. The necessary mechanical strength is provided by a relatively thick substrate of low resistivity silicon. By vapour deposition, the top surface of the substrate is covered with a thin layer of high resistivity silicon. If now the transistor is formed in this thin epitaxial layer, the required low capacitance and high resistivity is achieved, and the high series collector resistance is avoided. For the manufacture of a discrete transistor, the epitaxial layer has the same polarity as the substrate, i.e. n^+ upon n or p^+ upon p. For the formation of a transistor for use in an IC however, it will later be seen to be necessary for the epitaxial layer and the substrate to have opposite polarities. Thus, the device illustrated in Fig. 3.1 has a p-type substrate and an n-type layer, and the collector contact is necessarily brought out to the top surface.

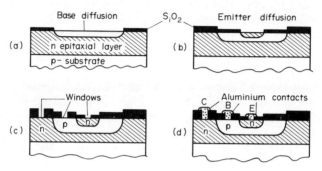

Fig. 3.1. Four stages in the formation of a transistor using epitaxial-planar techniques.

Surface oxidization

The important feature of the planar process is the deposition of a silicon dioxide layer on the top surface of the epitaxial wafer which acts as a mask against diffusion. The process involves exposing the wafer to an oxygen atmosphere at high temperature. In Fig. 3.1c it is seen that the oxide layer overlaps the exposed edges of the pn junctions. This protection from ambient conditions is such that, devices made using planar techniques are quite stable, even without encapsulation.

Photo resist and etching

After the oxidization process it is necessary to etch holes in the oxide, through which diffusion can take place. The process used is similar to that employed in the manufacture of printed circuit boards. Initially the oxidized surface is coated with a thin film of photo-sensitive emulsion (photo resist). A mask is manufactured, the pattern of which defines the area to be etched, it being opaque where etching is to be performed and transparent where the oxide is to be retained. The mask is brought into contact with the wafer and exposed to ultraviolet light. The photo resist under the transparent area of the mask, being subjected to this light, becomes polymerized and is not affected by the trichlorethylene developer which is subsequently used to dissolve the unexposed resist. When fixed, by baking, the remaining photo resist protects the oxide layer during the etching process. Hydrofluoric acid is used to remove oxide from the *window* where diffusion is required and, after the surface has been cleaned, the chip is ready for the first diffusion process.

Diffusion

For a p-type diffusion the most generally used dopant is boron. This is deposited on the wafer at high temperature, and diffuses through the window into the silicon. A p-type base region is thus created, as in Fig. 3.1a. The oxidization treatment is now repeated and, in this high-temperature process, the open window is sealed with an oxide layer and the base dopant is driven deeper into the silicon. A new mask is used in a second photo resist and etching stage, which opens a window for the diffusion of the emitter region (Fig. 3.1b). For n-type diffusion the most generally used dopants are phosphorous and arsenic. The cycle is repeated yet a third time. The emitter window is sealed by oxidization, the emitter dopant is driven in, and new windows are etched in the oxide layer to define the contact areas. This stage is illustrated in Fig. 3.1c. Finally the contacts are made by the evaporation of aluminium (Fig. 3.1d).

The foregoing explanation is, of necessity, much simplified and many more discrete processes are used than have been indicated. Typically, the manufacture of a bipolar transistor requires about 140 process steps. However, the essential features of manufacture are as stated. In practice many devices are manufactured at the same time on a single sheet of silicon. These are separated by scribing with a diamond stylus and breaking into individual chips. They are then mounted in suitable packages which allow electrical connections to be readily made and power, dissipated as heat, to escape.

3.2.   Bipolar integrated circuits[29]

Due to the differing behavior of boron and phosphorus in the manufacturing process, *npn* transistors are more readily made than *pnp* types. The latter require more specialized diffusion techniques, which explains the dominance of *npn* devices.

Integrated circuits are possible because, in addition to making transistors and diodes, the methods described may also be used to manufacture resistors and capacitors. After such devices have been formed, and isolated, windows for the necessary contacts are etched, as previously described. A vacuum deposition of aluminium is then made over the entire slice, and the photo-resist technique is used to remove unwanted material. The aluminium left provides the necessary contacts and interconnections between the various components which constitute the circuit.

Resistors

An IC resistor makes use of the fact that the more lightly doped is a piece of semiconductor material the greater is its resistivity. It most commonly takes the form of a thin *p*-type diffused strip with a contact at each end, the resistance value being proportional to the length of the strip and inversely proportional to its width. The range of values obtainable with diffused resistors is limited by the area of the chip which is available in a given circuit, but typically is between $10 \, \Omega$ and $30 \, k\Omega$, with a tolerance of not better than 20 per cent. An alternative method of producing resistors makes use of *thin-film* techniques. A metal film, usually nichrome, is deposited on the oxide layer and etched to obtain the desired shape, and hence resistance. This is covered with an insulating layer and ohmic contacts are made through the insulation. Thin film resistors make resistance values available up to $50 \, k\Omega$.

Capacitors

Where a capacitor is required in an IC, use is made of the depletion layer capacitance of a reverse biased $p^+ n$ junction, and the maximum value obtainable is about 400 pF. Twice this value can be obtained using thin-film techniques. An aluminium area on top of the oxide layer forms one side of a parallel plate capacitor, the other side being the semiconductor material, with the silicon dioxide layer acting as the dielectric.

Isolation

It is necessary to be able to electrically isolate individual devices from each other. This is done by surrounding each component with material of opposite polarity and reverse biasing the semiconductor junction so formed. This is illustrated in Fig. 3.2, which represents the cross-section of a bipolar IC containing an *npn* transistor and a resistor. Before the devices are formed, p^+ diffusions are carried

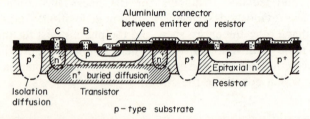

Fig. 3.2. An *npn* transistor and a resistor isolated by p^+ diffusions.

out, which are of sufficient depth to pass through the epitaxial layer and reach the *p*-type substrate below. This leaves isolated *n*-type pockets in which the various circuit components can be made.

An additional feature shown in Fig. 3.2 is the n^+ layer directly beneath the transistor. The discrete mesa transistor of Fig. 1.21 has its collector contact at the bottom of the device and the reason for using the epitaxial process was to reduce the resistance between the contact and the collector. In the IC structure shown, the collector contact is at the top of the chip and the "buried" n^+ layer, being in the path between the collector and the contact, serves to reduce series collector resistance in a similar manner. In the areas where it is intended to form transistors, n^+ *buried layer* diffusions are carried out on the *p*-type substrate before the epitaxial layer is grown. The dopant for these is usually arsenic. After epitaxy, isolated regions are created by deep p^+ diffusions, then the various components are formed, as previously described.

3.3.   Digital logic families[30]

From the very start, bipolar integrated circuits found wide use in digital logic systems. Logic devices provide a two-state output, for decision making, and control purposes, a high-voltage level (logic 1), or a low-voltage level (logic 0). These, in various applications, may be used to represent such alternatives as true–false, stop–go, yes–no, etc. In a typical logic system operating from $V_{CC} = +5\,\text{V}$, any voltage less than 0.4 V is interpreted as logic 0 and any voltage greater than 2.4 V is accepted as logic 1, although these limits vary between the various logic families.

As the field of application of logic devices widened, so did the requirement for circuits having low noise margins, high switching speeds, low power consumption, etc. Each logic family developed has shown some improvement over its predecessor. The merits of each are now discussed with reference to a simple gate circuit, which can be used to characterize their performance.

Resistor–transistor logic (RTL)

The circuit diagram of Fig. 3.3a represents a three-input NOT–OR (NOR) gate, the basic circuit of the Fairchild $RT\mu L$ range. It is described by the Boolean equation

$$f(x) = \overline{A + B + C}, \tag{3.1}$$

i.e. the output is NOT high when input A OR input B OR input C is high.

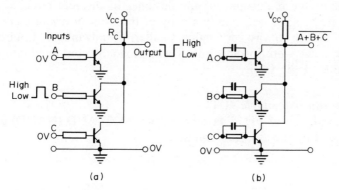

Fig. 3.3. Three-input NOR gates in (a) RTL and (b) RCTL.

With inputs A, B and C all at logic 0 the transistors are in the off-state. No current flows through the common collector resistor and the output is at V_{CC} volts (logic 1). If any input is taken to logic 1, the relevant transistor will go into saturation, and the resulting voltage drop across R_C will cause the output voltage to fall to nearly zero.

RTL is not basically a very fast configuration, but this Fairchild range represents a good compromise between speed and power consumption. Its noise margin is relatively poor, however, as is its fan-out capability.

The *d.c. noise margin* specifies the maximum value of a spurious signal which can be accommodated at an input, without causing the device output to change state. The *fan-out* capability indicates the maximum number of standard inputs, of devices in the same logic family, which can be driven from the device output. These IC parameters will be discussed in greater detail later.

Resistor–capacitor–transistor logic (RCTL)

A simple variation of RTL is the RCTL gate shown in Fig. 3.3b, representative of the Texas Inst. 51 series. In search of lower power consumption resistor values have been increased but, as a result, switching speeds are slower. Speed-up capacitors are included to compensate this.

A point of interest in these two circuits is that the IC designer was still using the philosophy of the designer of discrete circuits. An integrated circuit was simply a copy of the original discrete circuit and, although it was very much smaller, no advantages in circuit performance were forthcoming. This philosophy continued into the next main logic family.

Diode–transistor logic (DTL)

The diagram of Fig. 3.4a represents a NOT–AND (NAND) gate described by the Boolean equation

$$f(x) = \overline{A \cdot B \cdot C} \tag{3.2}$$

i.e. the output is NOT high when input A AND input B AND input C are high.

Referring to the circuit diagram, if any input is at logic 0 the relevant input diode conducts, and the bottom of R_1 is tied to approximately 0 V. The series diodes D_1 and D_2 do not conduct, the output transistor is cut off by the negative voltage through R_2, so its collector, and hence the output, stands at V_{CC} (logic 1). If *all* the inputs are now taken to logic 1, the input diodes are blocked and the bottom of R_1 rises from 0 V. This forward biases D_1 and D_2, forming a potential divider of R_1 and R_2 between V_{CC} and $-V$. Current now flows through R_1 to the base of the output transistor driving it into saturation. The resulting collector current causes a voltage drop across R_C and the output voltage falls to logic 0.

Note that when any input is at logic 0, current flows out through this input, via R_1, into the output of the preceding stage which drives it, i.e. into the collector of a saturated transistor. When a device is said to have a fan out of eight, say, this implies that the output transistor can *sink* currents from eight such sources, whilst maintaining the collector voltage below some stated value which defines the upper limit of logic 0.

One of the main advantages of DTL is that a number of outputs from different gates can be directly connected. This facility, known as *wired-OR*, can only be used where, when one gate is at logic 0 and another is at logic 1, the power supplies are not thus short circuited. Wired-OR can be used in other logic families but it was with DTL that it made its greatest impact.

In general, for low power dissipation, IC resistors are given high values whilst for high speed, low resistor values are necessary. The DTL family is characterized by a reasonable compromise between power dissipation and speed of switching, but has the disadvantage that an extra (negative) voltage supply is needed to assist in turning off the output transistor. This is overcome in the *modified* DTL gate of Fig. 3.4b, representative of the Texas series 15830 family. The series diode D_1 has been replaced by a transistor T_1, whose collector is taken to a "tapped" R_1 to prevent it going into saturation. This transistor provides an extra current source for driving the output

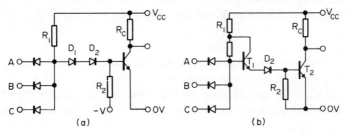

Fig. 3.4. Three-input NAND gates in (a) DTL and (b) modified DTL.

transistor T_2. As a result of this, and the lower value resistors used, higher switching speeds are obtained, and a gate propagation delay of 25 ns is typical.

The substitution of a transistor for a diode represented a change in design philosophy. It was recognized that in an integrated circuit, a transistor could be manufactured almost as cheaply as a diode. For instance, in the Series 53/73 logic family all diodes were replaced in this way, although the input transistor elements were still used essentially as diodes. As a result, in integrated circuits, active devices were used where, in discrete circuits, they would be precluded for reasons of economy. Consequently the performance of integrated circuits became superior to that of discrete circuits performing the same function.

Transistor–transistor logic (TTL)

Figure 3.5a illustrates the structure of a multiple emitter transistor used in TTL to replace the input diodes of DTL. It occupies less

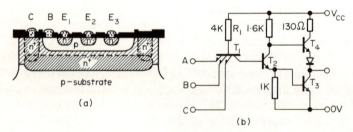

(a)

(b)

Fig. 3.5. (a) Structure of multiple emitter transistor as used in (b), a TTL NAND gate.

space than separate diodes, provides gain at the input and, in the *common base* input stage of the NAND gate of Fig. 3.5b, gives the device a low input impedance. As a result, source currents are larger, and faster operation is achieved.

In § 1.7 it was seen that a transistor could be represented by two diodes. The multiple emitter input stage can therefore be represented by the arrangement of Fig. 3.6a, in which the second series diode is the base–emitter junction of T_2. This transistor acts as a phase splitter. When cut off, its emitter is at zero volts and its collector is at V_{CC}. When it conducts, the emitter voltage rises and

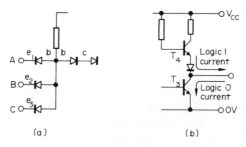

Fig. 3.6. (a) Diode representation of multiple emitter transistor. (b) Current flow in a "totem-pole" output stage.

the collector voltage falls. The emitter and collector of T_2 provide base current to transistors T_3 and T_4, respectively, which together form a push–pull (*totem pole*) output stage.

With inputs A, B and C all at logic 1, the emitter–base junction diodes of T_1 are reverse biased. The base–collector junction is forward biased and current flows from R_1 to the base of T_2, turning it on. The consequent rise in emitter voltage drives T_3 into saturation, the output falls to logic 0, and current flows from the load, through T_3 to earth, as in Fig. 3.6b. If any of A, B or C inputs are taken to logic 1, the relevant emitter–base diodes become forward biased and the current from R_1 flows through the input circuit into the source, instead of to the base of T_2. This transistor now cuts off, as does T_3, and T_4 conducts. The output now rises to logic 1, with current flowing *into* the load, again as shown in Fig. 3.6b.

Note that the outputs of several TTL gates cannot be connected for wired-OR operation. For if one gate was at logic 0 with T_3 conducting, and another was at logic 1 with T_4 conducting, V_{cc} would be effectively short circuited to earth.

In any circuit the time constants, associated with stray capacitance, tend to limit switching speed. An advantage of the totem-pole output is that the device output impedance is low in both the logic 0 and logic 1 output states. As a result, stray capacitances are quickly charged and discharged and *rise times* of the order of 5 to 10 ns are achieved.

The TTL family of digital circuits, as exemplified by the Texas 54/74 series, has proved to be a popular choice of systems engineers, and is made under licence by several manufacturers. Type 54 devices, with an operating temperature range of $-55°C$ to $+125°C$, have specific military and aerospace applications. The more generally used type 74 can be used in the range 0°C to 70°C. The family is further divided into groups having compatible performance ranges, as follows.

Standard. 54/74.   The circuit of Fig. 3.5b is representative of this type. It dissipates 10 mW per gate with a 50 per cent duty cycle and propagation delay time is typically 10 ns.

Low power. 54L/74L.   This uses the same circuit configuration as the standard series, but all resistors have higher values. As a result, the dissipation per gate is 1 mW, but the penalty paid is an increase in switching time to 33 ns.

High speed. 54H/74H.   For this the circuit has been modified. The output transistor T_4 is replaced by a *Darlington pair* arrangement which provides extra current drive for the logic 1 output state. This effectively reduces the charging time of load capacitance, and a 6 ns delay time is achieved. For this type, the dissipation is 22 mW per gate. In Fig. 3.7 note that diodes have been used to clamp the

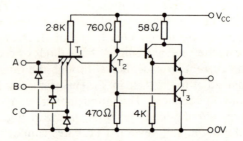

Fig. 3.7. Series 54H/74H (high-speed) NAND gate.

inputs to earth. These are included in the majority of devices and protect the input stage from damage due to the application of large spurious negative voltages.

Schottky. 54S/74S.   For ultra high speed it is necessary to reduce charge storage time, and in some devices, *gold doping* is used for this purpose. In the manufacturing process, gold is evaporated onto the back face of the silicon slice and diffused-in during the phosphorous emitter drive-in stage. The use of Schottky clamping diodes across the collector–base junctions of the relevant transistors provides an alternative method of achieving fast switching speeds.

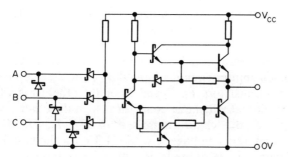

Fig. 3.8. Series 54S/74S (ultra-high-speed) Schottky diode NAND gate.

This is done in the 54S/74S series (Fig. 3.8), yielding a gate propagation delay time of 3 ns, with 19 mW per gate power dissipation. The diodes prevent the transistors going into heavy saturation, thus eliminating excess charge storage and speeding up recovery times. Aluminium is commonly used to form the diodes, at the same time as the interconnection layer is formed.

Emitter-coupled logic (ECL)

This is sometimes referred to as *common mode logic* (CML). Before the introduction of Schottky diode clamping, ECL provided the fastest switching circuits available. The basic gate, developed by Motorola is illustrated in Fig. 3.9. Transistors are operated in a non-saturated switching mode, and the problem of charge storage is thus avoided. In addition to its extremely fast switching capability

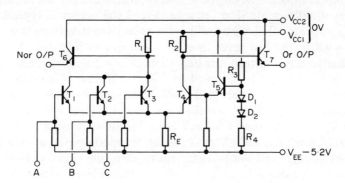

Fig. 3.9. OR/NOR gate using (ECL) emitter-coupled logic.

the ECL gate has the advantage that both the OR and NOR functions are obtained from the same device.

Basically the gate is comprised of an emitter-coupled differential amplifier, with emitter follower outputs taken from each collector. The common emitter circuit is formed by T_4 on the one side, and any of the input transistors T_1, T_2 and T_3 on the other. It is a feature of such a circuit that when one collector is high the other is low. With $V_{EE} = -5.2$ V, the chain comprised of R_3, T_5, D_1, D_2 and R_4 provides a reference voltage V_{BB} of -1.3 V for the base bias of T_4. If any input voltage exceeds V_{BB} that input transistor conducts, voltage is dropped across R_1, and a corresponding voltage drop occurs at the NOR output. If all inputs are now taken to logic 0 (less than V_{BB}) the input transistors cut off and the common emitter voltage falls. This causes T_4 to conduct, voltage is dropped across R_2 and a corresponding voltage drop occurs at the OR output.

In the MECL family, as a device switches from logic 0 to logic 1, the output voltage changes from about -1.7 V to -0.9 V, a voltage swing of 0.8 V.

3.4. Noise immunity

Integrated circuits have tended, in general, to work at lower voltage levels than discrete component circuits, and this fact explains why users and manufacturers are concerned with the noise immunity of different types of circuits. It is a relatively easy matter to say what is the maximum noise voltage which can be tolerated at any input, without causing false operation, and this is what manufacturers usually quote. For instance, Fig. 3.10a, which is typical for TTL, indicates that a gate will react to any input voltage less than $V_{IL} = 0.8$ V (the low level threshold voltage) as if it were logic 0. Similarly, any voltage greater than $V_{IH} = 2$ V (the high level threshold voltage) is accepted as logic 1. If now a manufacturer guarantees that for a specified loading, the logic 0 output signal will not exceed $V_{OL} = 0.4$ V, and the logic 1 output signal will not be less than $V_{OH} = 2.4$ V, then a 400-mV noise immunity is implied. A noise signal of less than 400 mV peak, which is superimposed upon a gate output voltage, will not be sufficient to carry the signal beyond the relevant threshold level and cause indeterminate action of the device to which it is applied.

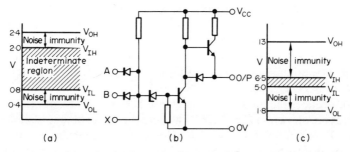

Fig. 3.10. Comparison of noise immunity for (a) series 74 and (c) series 74 HiNL, devices.

Noise signals are most frequently generated by stray capacitive coupling between output and input leads, and the voltage generated at an input, by a given disturbance acting through a stray capacitance, depends very much upon the impedance of the input circuit. A high noise immunity coupled with a low impedance is therefore more valuable than when coupled with a high impedance. In view of this it is dangerous to make generalizations about the relative noise immunity of different classes of logic and each manufacturer's products must be separately examined from this point of view.

To mitigate noise problems, the use of a ground plane in close proximity to the wiring plane is recommended by some manufacturers and careful attention must be paid to power earthing and to decoupling of power leads. This is discussed in the final chapter of this book.

High noise-immunity logic (HNIL)

Reference to Fig. 3.10a indicates that the basic noise immunity problem is the narrow spread between output voltage levels and input logic threshold voltages. The logic families so far discussed have been developed in search of high speed combined with low-power dissipation and, where these have been used in a noisy environment, highly regulated power supplies and screening arrangements have been necessary. However, for most applications outside the computer field, high speed is not usually required and, if the degree of regulation can be relaxed, power supplies need not be expensive. Accordingly, a family of logic devices has been developed which, by operating with higher voltage levels, gives greater noise immunity, but this is achieved at the expense of speed and power. These devices have different styling according to their manufacturer, for instance, Texas *HNIL* and Teledyne *HiNIL* (high noise-immunity logic), S.G.S. *HLL* (high level logic), and Motorola *HTL* (high threshold logic).

Reference is made to the typical HiNL NAND gate, developed by Teledyne, shown in Fig. 3.10b. The general arrangement is similar to the DTL circuit of Fig. 3.4b. The effect of the zener diode is to raise the input threshold levels by 5.8 V and, for a device operating from $V_{CC} = +15$ V ± 1 V, the noise immunity is as indicated in Fig. 3.10c. A point of interest in this circuit is the input marked X. A number of external diodes can be connected to this point and the *fan in* of the device thus increased.

3.5. Summary of bipolar digital circuits

Attention has so far been focused on logic devices as an introduction to various types of bipolar integrated circuits, and their characteristics, together with those of MOST devices, are summarized in Table 3.1.

The gates described, in addition to being used independently in logic systems, form the basic elements from which more complex integrated circuits are made.

Typical of such devices are decade counters, multivibrators, shift registers, etc. and these are discussed in subsequent chapters.

Table 3.1. Typical characteristics of digital logic families

Family	Dissipation, (mW/gate)	Propagation delay (nS)	Noise immunity (V)	Fan out	Logic swing (V)
DTL	5	25	0.5	8	2.1
TTL 74	10	10	0.4	10	3.3
74L	1	33	0.4	10	3.3
74H	22	6	0.4	10	3.3
74S	19	3	0.4	10	3.3
HNIL	60	110	6.5	10	14.0
ECL	25	2	0.27	25	0.8
C-MOS	10 nW	25	45% of V_{DD}	> 50	Up to 15

Circuit types which depend on bipolar technology have, however, been limited in complexity because of yield problems in manufacture. These problems are being overcome and new types are being developed with the promise of high manufacturing yield. Of these, the most promising is *collector diffusion isolation* (CDI) which allows high-speed operation of systems having in excess of 7000 components on a chip area less than $0.4\,cm^2$.

3.6. Linear circuits[31]

Because so many digital circuits made use of transistors operating in a saturated mode, large-scale production of these devices was possible, in spite of the limitations to capacitor and resistor values which could be obtained. Because of these limiting values, however, together with poor tolerances and temperature coefficients, bipolar linear integrated circuits developed much later than digital types.

It was with the adoption of the new design philosophy, in which passive components are replaced by active components, that these problems were solved. Linear circuits were then produced which were more complex than equivalent discrete circuits, and having excellent performance characteristics. An early outstanding example was the μA 709 *operational amplifier* designed by R. J. Widlar, containing fourteen transistors and fifteen resistors. It provided a performance at least equal to that of a discrete circuit but at far lower cost. However, it needed additional external components, to obtain stable operation and has now been replaced by the widely used μA 741, which has internal frequency compensation. This low-cost amplifier has a slew rate of $0.5\,V/\mu s$ (the maximum rate of change of output voltage it can produce) and an input impedance of $1\,M\Omega$. Development in the operational amplifier field has aimed at improving such specific characteristics as bandwidth, output power, slew rate and input impedance, etc. As examples, there are available, high-voltage ($\pm 40\,V$) amplifiers capable of delivering 1 A to a load, the μA 715 with a slew rate of $18\,V/\mu s$, and the 8007 FET input amplifier providing an input impedance of $10^6\,M\Omega$.

A second class of circuit which has found wide use has been the *voltage regulator*, used for the regulation of power supplies for integrated circuit systems. A typical example of this is the LM 109 providing an output of up to 1.5 A with an output impedance of $30\,m\Omega$.

Other circuits now being produced in integrated form include, i.f. and r.f. amplifiers, analogue multipliers and phase-locked loops, and such devices are considered in the relevant chapters of this book.

3.7. MOS integrated circuits

The number of discrete steps in the manufacture of an MOS transistor is about 35 compared with some 140 steps which are needed to make a bipolar device. This reduction in the number of process steps is accompanied by a higher production yield and hence lower costs. Reference to Fig. 2.41 shows that only one diffusion process is used and, when the MOST is included in an integrated circuit, no isolation diffusions are necessary. This is because the junctions, formed by the drain and source p diffusions in the n-type substrate, are reverse biased by the potentials normally applied when the device is in use. Combined with the fact that one device can be directly coupled into the gate of the next, this means that MOS circuits are very much smaller than equivalent bipolar units. Alternatively, many more MOS devices can be accommodated on a given area of silicon.

Just as, in bipolar technology, it was easier to make an npn transistor than a pnp, so in MOS integrated circuits, the more easily made transistor was the p-channel type using an n-type substrate. It will be recalled that the application of a negative potential at the gate induces positive charge in the underlying silicon, to form the conduction channel. The minimum voltage which causes conduction is V_T, the threshold voltage, and in early devices this was 4 or 5 V. For satisfactory operation a power supply of some five times V_T was necessary, and the required higher operating voltages made them incompatible with bipolar devices. When MOS and TTL integrated circuits were used in the same system, special *interface* circuits were necessary. Accordingly, development was aimed at the reduction of V_T.

Metal–nitride–oxide–semiconductor (MNOS)

The threshold voltage of an MOS transistor is proportional to the work function between the metal gate and the silicon, to the thickness of the gate dielectric, and is inversely proportional to the dielectric constant of the insulator. Silicon nitride has a dielectric constant twice that of silicon dioxide but, in production, the interface between nitride and silicon is difficult to control. The problem is overcome in the MNOS structure illustrated in Fig. 3.11a.

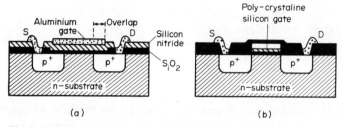

Fig. 3.11. Structure of a transistor in (a) MNOS and (b) silicon gate forms.

The gate dielectric consists of a thin silicon dioxide layer, next to the silicon surface, with a relatively thick silicon nitride layer immediately above it. Such a device has a threshold voltage of about 2 V, and the interface between the two dielectric layers is capable of storing charge for very long periods of time. This latter feature makes the MNOS suitable for use in shift registers and non-volatile memories. A typical production range is styled GIANT (General Inst. advanced nitride technology), and is compatible with TTL devices.

Silicon gate MOS

The gate electrode of a *metal gate* MOST is made larger than is theoretically necessary in order to allow for mask alignment errors during manufacture, and some overlap of source and drain occurs. As a result, inter-electrode capacitances are formed and these set a limit to the switching speed of the device, of about 1 MHz. In the *silicon gate* transistor of Fig. 3.11b, the gate electrode is formed with *p*-doped polycrystaline silicon. This has a lower work function than the metal gate and yields a threshold voltage compatible with that of MNOS.

In the manufacturing process, the gate is deposited first, and then used as a mask for the *p* diffusion of source and drain. The technique is said to be *self-aligning*. The very small overlap which occurs is due to sideways diffusion and, as a result of the reduced interelectrode capacitance, switching speeds up to about 15 MHz can be achieved.

A silicon gate transistor can be made much smaller than an equivalent metal gate device; a further advantage arises from its method of manufacture. The gate oxide is covered with silicon as soon as it is grown, and can then withstand additional high-temperature oxidization and diffusion processes. Thus, it can be included in *hybrid circuits* requiring both MOS and bipolar devices on the same silicon slice.

Some other techniques

Different solutions continue to emerge to the problem of providing more circuitry, in a smaller space, operating at higher frequencies. These include, SATO (self-aligned thick oxide), and RMO (refractory metal oxide), both of which are self-aligning techniques. Silicon-on-sapphire (SOS) is another process which aims at reducing stray capacitance. In this, devices are formed in silicon wells sunk in a sapphire substrate. A process of significance, used to minimize gate overlap in metal gate fabrication, is that of *ion implantation.* Initially the metal gate is made shorter than the channel length between the source and drain. A high energy beam of boron ions is then used on the slice. The ions penetrate the thin oxide and enter the silicon, extending both the source and drain *p* regions up to the edges of the gate. The metal gate and dielectric, however, absorb the ions and no penetration of the underlying silicon occurs. Use of the implantation process results in practically no gate overlap, and such devices can operate up to about 20 MHz.

Of the many manufacturing processes used in MOST technology, the aluminium gate and the silicon gate techniques are two which have become accepted as standard. Each process can now be manufactured with single polarity transistors, either as *p* channel (PMOST) or as *n* channel (NMOST). Alternatively both polarity types can be combined on a single chip to form complementary MOS devices.

3.8.   Complementary MOS

In bipolar circuits, resistors are formed during base diffusion and a practical limit of about 30 kΩ has been indicated. In MOS

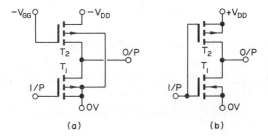

Fig. 3.12. Comparison of (a) P-MOST and (b) C-MOST inverters.

technology a transistor can be used as an active load resistor. Since the on-resistance of the device is the reciprocal of its transconductance, a high value resistor can be obtained by forming a low y_{fs} transistor. In the PMOST inverter circuit of Fig. 3.12a, T_2 is the load resistance; the gate is sometimes connected to the drain for this purpose. Let it represent a load of 100 kΩ, and let the on-resistance of T_1 be 10 kΩ. If with a V_{DD} of -20 V V_T is -3 V, then a negative voltage at the input, greater than this, will cause T_1 to conduct. With T_2, this acts as a potential divider and the output rises from -20 V to -1.8 V.

This is compared with the inverter circuit of Fig. 3.12b which is the basic element of C-MOS devices. When the input is at zero volts (logic 0), the n-channel transistor T_1 is off and p channel T_2 is on. The output is thus essentially shorted to the positive voltage rail V_{DD}. No current flows from the supply to earth via the inverter and, since the output terminal feeds a high input impedance device (another MOST), logic 1 current drawn from T_2 is minimal. When the input is taken to logic 1, T_1 turns on and T_2 turns off. Now the output is shorted to earth, and again no current flows through the inverter. Thus, C-MOS devices dissipate very little power and their logic voltage swing is over almost the full range from zero volts to V_{DD}. Accordingly their noise immunity is comparable to that of HNIL devices.

Structure and characteristics

Either metal gate or silicon gate techniques can be used in manufacture; the structure of the former type is illustrated in Fig. 3.13. It is made by first preparing the n-type substrate with a p

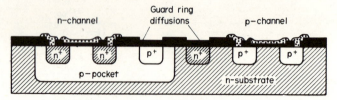

Fig. 3.13. Complementary MOST structure.

diffused pocket in which the n-channel transistor is formed. The p-channel device is formed in the undisturbed substrate as before. To simplify stability problems C-MOS circuits are arranged so that the sources of n-channel transistors are held at earth potential.

Complementary MOS devices are characterized by a single power-supply requirement (3 V to 15 V), minimal power dissipation, high noise immunity and a high fan-out capability. They are compatible with DTL and TTL logic families and will compete with these in medium speed applications up to about 25 MHz. Their characteristics are compared with those of other logic families in Table 3.1.

3.9.   Charge-coupled devices[32]

In § 2.10 it was seen that in an MOS device, a capacitor was formed between the metallic gate and the silicon substrate, with the oxide as dielectric. The earliest form of charge-coupled device (CCD) was thought of as an array of such capacitors, closely spaced so that charge could be transferred from one capacitor to the next. In this respect a CCD structure is similar to a *shift register* (Fig. 9.25b) in which, under the control of a train of "clock" pulses, binary digits can be caused to move along a line of bistables connected in series. The CCD has the advantage, however, that transferred information may be either analogue or digital in form.

In a shift register, charges, representing signal information, can be introduced by the application of electric potentials. It is an additional feature of charge-coupled devices that charges can be generated by the application of light, the magnitude of charge being proportional to the light intensity. Since such charges can be moved through a CCD structure to some output terminal, the device is capable of being operated as an *image detector*. The two methods of introducing charge are illustrated in Fig. 3.14.

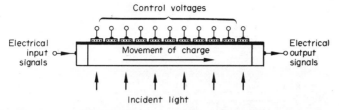

Fig. 3.14. CCD shift register. Charge may be introduced, either by the application of electric potential or by incident light.

Principle of action

Referring to Fig. 2.42, the application of a negative potential at the gate induces positive charge in the underlying silicon substrate, and a depletion layer is formed. With a continuing increase in this negative bias, the layer penetrates deeper into the substrate. At some bias threshold voltage V_T, the interface between the silicon and oxide becomes so negative that holes are attracted to the surface to form a thin "inversion layer", as indicated in Fig. 3.15b.

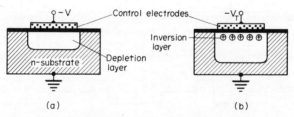

Fig. 3.15. Elementary CCD structure. At the threshold voltage $-V_T$, a "potential well" is formed in which charge can be stored.

Thus there exists a "potential well" in which information can be stored in the form of varying amounts of positive charge.

Given that such a charge exists in one element, if the "gate" (or control electrode) of an adjacent element is subjected to a more negative control potential, the charge will move from the first into the second along the surface of the substrate. If every third control electrode is connected, as in Fig. 3.16, then the application of a sequence of suitable bias voltages will cause a charge, introduced at the input on the left, to be transferred step by step until it appears at the output on the right. The sequence of one shift is illustrated in

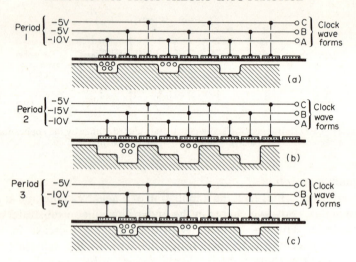

Fig. 3.16. Three-phase clocking of a charge-coupled device.

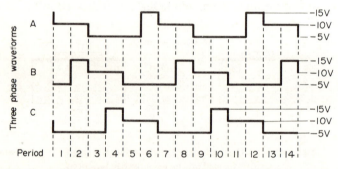

Fig. 3.17. Three-phase waveforms suitable for clocking the shift register of Fig. 3.16.

Fig. 3.16 by considering (a), (b) and (c) in succession. The control signals for these three states are those shown in periods 1, 2 and 3 of the waveforms of Fig. 3.17, and the application of these three trains of signals would cause sequential operation of the shift register.

A practical shift register[33]

Figure 3.18 illustrates a practical form of CCD shift register. Minority carriers (holes), from the forward biased input diode D_1, pass through a conduction channel induced by the input "gate", and

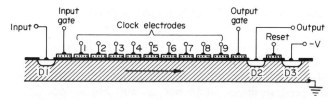

Fig. 3.18. Input–output arrangements for a CCD register.

so reach the first CCD element. At the output, the signal is retrieved from diode D_2 which is reverse biased so as to act as a sink for the holes received from the last stage. The function of D_3, which is held at a more negative potential, is to reset the D_2 potential after it has received a charge packet. This is done by momentarily connecting D_3 to D_2 by the induced channel of the reset electrode.

Transit time limitation

The CCD is, in effect, a *dynamic memory* and stored information disappears with the passing of time. This is due to the thermal generation of electron–hole pairs (called the dark current), a phenomenon which is common to all semiconductor devices. Because of this effect the depletion layer slowly fills with minority carriers and stored information is gradually masked. It is thus necessary for any charge packet to pass through a CCD structure in a time so short that this extra thermally generated charge is negligible; in other words there is a maximum allowable signal transit time. However, this is typically many milliseconds, and since clock rates of at least 1 MHz are normally used, the limit does not place too great a restriction on shift register length.

Optical applications [34]

In § 1.5 it was shown that, when light falls on silicon, minority carriers are generated and in a CCD structure some of these will flow into the potential wells. Thus, charges are introduced which are proportional to the light intensity. If time is allowed for these to accumulate (1 ms is typical), and then, by normal shift register action such charges are transmitted to the output, electrical signals are obtained which are proportional to the varying light intensity along a line of CCD elements. A matrix of such elements, forming many lines one above the other, can therefore provide electrical signals which describe a two-dimensional image, in other words it will perform the function of a television camera. All that is necessary to build up an image is that a line should contain enough elements for optical purposes, and that each line should be scanned in some predetermined sequence at suitable speed. In the *line transfer* system of Fig. 3.19, the address circuitry selects each line in sequence, and signals are fed to the output via a shift register and high input impedance amplifier. This is described in detail, and compared with the *interline transfer* and the *frame transfer* systems, in reference 34.

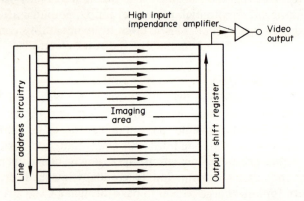

Fig. 3.19.   Line transfer organization for a CCD area imager.

Development

A description has here been given of an early charge-coupled device. As with any other emergent semiconductor technology, however, continuing development has produced variations on the basic concept and a designer has various options open to him. Thus, there are available p-channel and n-channel devices requiring two-, three- or four-phase clocking. A three-phase device can only be shifted in one direction, while movement is possible in both directions with the two-phase device.

An alternative to the *surface channel* element is that which employs the *buried channel* mode of operation. In this, the doping of the silicon substrate is modified such that both storage and transfer of charge takes place in the bulk silicon, just beneath its surface. In addition to simplifying device design, the buried channel type provides several major advantages in performance. In particular, charge transfer efficiency is high for the full range of charge packet size, although a criticism of the buried channel mode is that the *saturation* charge density cannot be made as high as with the surface channel mode, which limits its dynamic range. However, such limitation is more than compensated by the relatively low noise levels which have been obtained.

Applications [35]

The range of possible uses of charge-coupled devices is a wide one, and each application has its own particular requirements. Thus, for a standard television array as a replacement for the vidicon in television cameras, there are required large CCD chips. For low-level light imaging arrays a good signal-to-noise ratio is necessary, while for infrared detection, dark current must be minimal.

Apart from the optical field, suggested uses for charge-coupled devices include dynamic memories, analogue time delays, multiplexing, transverse filtering and recursive filtering. For such applications the design emphasis will probably be on achieving wide dynamic range, high transfer efficiency and high packing density. Much research is being devoted to the CCD in attempts to satisfy these potentially large markets.

CHAPTER 4

Amplifiers

INTRODUCTION

It is the purpose of an electronic amplifier to accept very small signals at its input, amplify them, and then to use the resulting magnified signals to perform some given function. For instance, the input signals for an audio amplifier are obtained from a microphone, the pick-up of a record player, or the playback head of a tape recorder and, when amplified, are used to provide the current drive for a loudspeaker. Such an application requires amplification over a frequency range of about 30 Hz to 20 kHz, with a minimum amount of distortion. By comparison, amplifiers for use in servo systems and analogue computers are required to amplify down to zero frequency. In the specialist field of medical electronics, amplified signals obtained from the brain, heart or nerves are used for display and recording purposes and, due to the very small signals available, amplifiers are required to combine high gain with low noise.

The general approach to amplifier design is first to deal with the output stage. Then, knowing the level of signals required to drive it, to design the preceding stage accordingly, and thus to work back from the output of the amplifier to its input. The subject will be treated in that order.

4.1. Power amplifiers

In a power amplifier the object is to obtain the maximum amount of power, while limiting distortion to a predetermined level. The transistor is particularly well suited for use in power amplifiers. It is linear over practically the whole of its collector characteristic, and is so efficient that it is almost possible to achieve the maximum theoretical efficiency of 50 per cent for Class A and 78.5 per cent for Class B amplification. The common emitter configuration is most widely used since it provides much greater power gain than the common collector arrangement. Its inherent distortion level is higher, but this can be largely compensated by the use of negative feedback.

The class A power amplifier

The main steps in the design of such an amplifier are, the choice of operating point such that operation is restricted to the linear part of the characteristic, the biasing arrangement to obtain this operating point, and the determination of the correct load.

Design considerations

In a power amplifier there is very real danger of thermal runaway unless the biasing arrangement chosen is such as to prevent shift of the d.c. operating point. The circuit of Fig. 4.1 is widely used for this

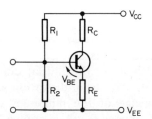

Fig. 4.1. Basic biasing arrangement of a transistor power amplifier stage.

purpose and will provide an acceptable degree of stability. The emitter resistor R_E introduces voltage negative feedback and, in conjunction with the base resistors R_1 and R_2, determines the value of the input voltage V_{BE}. An increase in emitter current causes a voltage drop across R_E and reduces the base–emitter voltage. The base current is thus reduced, providing a large degree of compensation for the original change. If a stability factor is defined as $K = \delta I_{CEO}/\delta I_C$ (as in § 1.15), minimum stability is obtained when K is unity and the greater the value of K the better the stability becomes. For the circuit of Fig. 4.1, if R_E is large compared with R_B, this factor is

$$K = 1 + \frac{\beta R_E}{R_E + R_B}, \tag{1.26}$$

where $R_B = R_1R_2/(R_1 + R_2)$.

The equation shows that the greater the value of R_E and the smaller the value of R_B, the higher will be the value of K, and hence the better the stability. The maximum value of R_E depends on how much of the supply voltage can be dropped across it, and so how much voltage is available as signal swing in the collector resistor. Similarly, the minimum values for R_1 and R_2 depend on how much current may be drawn from the power supply by them. At very low values this potential divider will shunt the a.c. input. As a first step it is reasonable to make R_B equal to $10R_E$ and then, using eqn. (1.26), to check the degree of stability obtained. In the design of this circuit, two other equations are required. Referring to Fig. 4.1,

$$V_B = V_{CC}R_2/(R_1 + R_2). \tag{4.1}$$

Also,

$$V_{BE} = V_B - I_ER_E, \tag{4.2}$$

from which

$$V_B = I_ER_E + V_{BE}. \tag{4.3}$$

Design steps
1. Select a suitable transistor.
2. Plot the load line on the output characteristic, select an operating point, and check distortion and power output.
3. With a knowledge of the minimum voltage to which the collector falls, choose a suitable value for R_E, to make V_E a little less than $V_{C(min)}$.
4. Using eqn. (1.26), determine the value of R_B for the required stability factor.
5. Using eqns. (4.1) and (4.3) calculate the values of R_1 and R_2.
6. Select a suitable capacitor to decouple R_E.

DESIGN EXAMPLE 4.1

Required, an a.c. power of 2 W into a resistive load of 10 Ω, with low distortion and a stability factor of 8. A supply of 24 V is available.

To obtain an output with low distortion, a transistor is selected having a much greater power rating than the power actually required. Let it be a BD124 having a rating of 15 W at temperatures below 60°C, and whose output characteristic is given in Fig. 4.2.

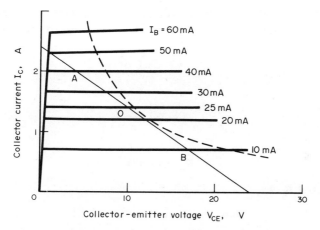

Fig. 4.2. Characteristic curves of a BD 124 transistor.

Load line. With the load line as drawn, the operating point chosen is $V_C = 10$ V, $I_C = 1.4$ A, corresponding to an I_B of 25 mA. For an input signal of 30 mA peak to peak, the output voltage varies between 4 and 17 V as the current swings through 1.3 A.

The power developed in the load is given by the expression

$$\text{Power} = \frac{(V_{\max} - V_{\min})(I_{\max} - I_{\min})}{8}. \tag{4.4}$$

Thus, a power of 2.1 W is obtained.

The percentage harmonic distortion is given by the expression

$$\frac{(I_{\max} + I_{\min} - 2I_0) \times 100}{2(I_{\max} - I_{\min})} \tag{4.5}$$

However, a quick assessment of distortion may be made by measuring the load line and calculating the ratio of its two parts. For instance, in Fig. 4.2, with the operating point at O, if the ratio of BO to AO is 11 to 9, i.e. a ratio of 1.22, then 5 per cent second harmonic distortion is present. If this figure is less than 1.22, the distortion is less than 5 per cent, there being no distortion when the two parts are equal. For the load line drawn the ratio $BO : AO = 1.17$ indicating an acceptable distortion level.

Emitter resistor. Since $V_{C(\min)} = 4$ V and the standing collector current is 1.4 A, let $R_E = 2.2\ \Omega$.

Stability factor.   The transistor manufacturer gives $\beta = 50$ for an I_C of 1.5 A. Substituting β and R_E in eqn. (1.26), for $K = 8$,

$$8 = 1 + \frac{50 \times 2.2}{2.2 + R_B},$$

therefore,

$$7R_B = (50 \times 2.2) - 15.4,$$

so,

$$R_B = 13.5 \ \Omega.$$

R_1 and R_2.

$$V_E = I_E R_E \doteqdot V_B \quad \text{[from eqn. (4.3)]}.$$

But,

$$V_B = \frac{V_{CC}R_2}{R_1 + R_2} \quad \text{[from eqn. (4.1)]},$$

therefore,

$$3.08 = 24R_2/(R_1 + R_2),$$

from which

$$R_1 \doteqdot 7R_2$$
$$R_B = \frac{R_1 R_2}{R_1 + R_2},$$

therefore

$$13.5 = \frac{7R_2{}^2}{8R_2} \quad \text{and} \quad R_2 = 15.4 \ \Omega.$$

Therefore

$$R_1 = 7R_2 = 107.8 \ \Omega. \tag{4.6}$$

Let the values used be $R_1 = 110 \ \Omega$ and $R_2 = 15 \ \Omega$. The current drain due to R_1 and R_2 in series is $24 \ \text{V}/125 \ \Omega \doteqdot 200 \ \text{mA}$ which is small compared with the standing current of the transistor.

The final circuit values are therefore, $R_C = 100 \ \Omega$, $R_1 = 110 \ \Omega$, $R_2 = 15 \ \Omega$, and $R_E = 2.2 \ \Omega$.

Decoupling capacitor.   At the minimum frequency at which the stage is to be used, the reactance of the decoupling capacitor should be small compared with 2.2 Ω. For instance, if the minimum frequency is 1 kHz, then at this frequency a 1000 μF capacitor has a reactance of about one-sixth of the emitter resistor which is satisfactory.

4.2.   Audio power amplifier, class A

Design considerations

The loudspeaker is coupled to the transistor by an output transformer and if the operating point is at the mid-point of the dynamic collector loadline, then it is possible for the collector voltage to swing between the knee voltage and twice the supply voltage, as shown in Fig. 4.3. The collector current swings from zero to twice

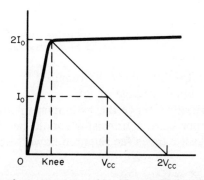

Fig. 4.3. Calculation of power output and load from a single collector characteristic. Note that for the purpose of illustration V_{knee}, normally less than 1 V, has been exaggerated with respect to V_{CC} (see Fig. 4.4).

the standing current. The load represented by this load line is therefore:

$$R_L = (V_{CC} - V_{\text{knee}})/I_o. \qquad (4.7)$$

The maximum power output obtainable from this load is:

$$P_{o\,(\text{max})} = \frac{(V_{CC} - V_{\text{knee}})}{\sqrt{2}} \cdot \frac{I_o}{\sqrt{2}}. \qquad (4.8)$$

Combining these two equations,

$$R_L = \frac{(V_{CC} - V_{\text{knee}})^2}{2P_{o(\text{max})}}. \tag{4.9}$$

These relationships may be used in the design of a power amplifier where a high degree of efficiency is required.

Design steps
1. Allow for a slightly higher power than is actually required and, using eqn. (4.9), evaluate the required load R_L.
2. Determine the turns ratio of the transformer to match the speaker to this load.
3. Note the lowest voltage to which the collector swings and choose R_E such that the voltage dropped across it is less than this voltage.
4. Making $R_B = 10R_E$, evaluate R_1 and R_2.
5. Choose a suitable capacitor with which to decouple R_E.

DESIGN EXAMPLE 4.2

A maximum output of 60 mW is required for a portable radio receiver using a supply voltage of 9 V.

Choice of transistor. Since the maximum theoretical efficiency obtainable is 50 per cent, a transistor should be chosen having a power rating at least 3 times the required power. Let it be a 2N 526

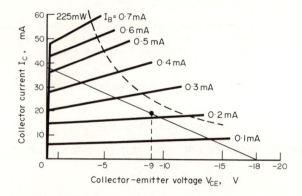

Fig. 4.4. Collector characteristic of a 2N 526.

having a rating of 225 mW at 25°C. From the collector characteristic of Fig. 4.4, the knee voltage is seen to be about 0.25 V.

Choice of load. In making use of eqn. (4.9) to evaluate the required load, allow for a power output of 70 mW. This will take account of loss in the ohmic resistance of the transformer primary, and also allow for the fact that the collector current cannot be taken down to zero because of leakage. In addition, to ensure that distortion is a minimum, consider the knee voltage to be 1.0 V. Then,

$$R_L = (9 - 1)^2/(2 \times 0.070) = 457 \ \Omega.$$

Transformation ratio. The transformer required to match a 3 Ω speaker to this load should have a turns ratio of $\sqrt{(457/3)} = 12.3:1$.

Emitter resistor R_E. A load line for 457 Ω is now plotted on the output characteristic from the point $I_C = 0$, $V_C = 18$ V, and the current at the operating point, $V_C = 9$ V, is seen to be 19.5 mA. If the current is allowed to swing from 36 mA down to 3 mA, while the collector voltage swings from -1.5 to -16.5 V, an output of 62 mW is obtained. The emitter voltage may therefore be made 1.0 V. Thus, $R_E = 1$ V/19.5 mA $= 51 \ \Omega$, a preferred value.

Now make $R_B = 510 \ \Omega$ (i.e. $10R_E$),

$$V_B = R_2 V_{CC}/(R_1 + R_2) \doteqdot V_E.$$

Therefore

$$1.0 = 9R_2/(R_1 + R_2) \quad \text{and} \quad R_1 = 8R_2,$$

but

$$R_B = R_1 R_2/(R_1 + R_2),$$

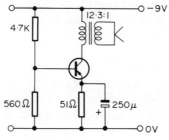

Fig. 4.5. Completed circuit of Design Example 4.2.

therefore

$$510 = 8R_2^2/9R_2.$$

This yields $R_2 = 574\,\Omega$ and $R_1 = 4592\,\Omega$, the nearest preferred values being $560\,\Omega$ and $4.7\,\text{k}\Omega$. To complete the design a capacitor is chosen to decouple R_E. The reactance of a $250\text{-}\mu\text{F}$ capacitor at $60\,\text{Hz}$ is approximately $10\,\Omega$, which is satisfactory. The final design is therefore as shown in Fig. 4.5.

4.3.   The class B push–pull amplifier

In a push–pull amplifier the two transistors are excited by equal base signals, 180° out of phase, and the outputs are combined by means of a centre tapped transformer. Assuming identical transistors, the push–pull arrangement has the following advantages over a single transistor stage:

1. No signal frequency currents flow in the voltage-supply source and hence there is no feedback to previous stages using the same supply.
2. Even order harmonics cancel out, thus providing less distortion for a given amount of power.

Design considerations

Theoretically, a Class B push–pull amplifier should have its two transistors biased to cut-off, but in practice this causes *cross-over distortion* if the change over in current from one transistor to the other is not smooth. This type of distortion may be largely overcome by supplying the drive for the stage from a high-resistance source and by applying a small forward bias to each transistor. The bias would typically be 100–200 mV giving rise to a quiescent current of a few milliamps. In the basic circuit of Fig. 4.6 the bias is provided by R_1 and R_2.

The value of V_{BE} required at a transistor for any given collector current falls as the temperature rises, a decrease of 2.5 mV per °C being typical. The temperature range over which the stage is to be used should therefore be considered since, with an increase in temperature, I_o rises and may reach such a magnitude that, in spite of the fixed bias provided, cross-over distortion is again present.

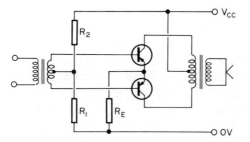

Fig. 4.6. Basic circuit of a Class B push–pull amplifier.

Similarly, a large reduction in ambient temperature may so reduce the quiescent current that it becomes insufficient to reduce this distortion.

Such effects of temperature changes may be minimized by shunting R_2 with a negative temperature coefficient thermistor. Thus, as temperature rises, the resistance of the parallel combination falls and the base voltage V_{BE} is decreased, offsetting the rise in collector current. A resistor in the emitter circuit R_E similarly increases stability but, as this is at the expense of efficiency, it is usually of a low value.

The amount of power dissipated by the stage is dependent on the amplitude of the incoming signal and under no-signal conditions very little current is drawn from the supply source. The peak output power which the circuit can handle is a little less than 5 times the maximum collector dissipation of each transistor, and efficiencies of 70 per cent may be readily achieved.

Assuming ideal transistors, the load presented to each collector is:

$$R_C = V_{cc}^2 / 2P_{o(max)}. \qquad (4.10)$$

However, in using this equation, a voltage somewhat less than V_{CC} should be considered, allowing say 0.5 V to avoid distortion as the bottoming voltage is approached. Similarly, the value for $P_{o(max)}$ should be higher than that required since the load on the transistor includes the unbypassed resistor R_E and some power is lost in this resistor. The useful power output is given by

$$P_{eff} = P_{o(max)} \cdot R_C / (R_C + R_E). \qquad (4.11)$$

Design steps

1. Select a suitable pair of matched transistors.
2. Using eqn. (4.10) evaluate the load to be presented to each collector and determine the ratio of the output transformer.
3. Select suitable values for R_1, R_2 and R_E and check that, with the value of R_E, the output power to the transformer is sufficient.

DESIGN EXAMPLE 4.3

Required, a Class B push–pull output stage giving a peak power output of 1.25 W. A supply of 15 V is available.

Transistors. The selected transistors must be capable of dissipating one-fifth of the peak power output. A matched pair of devices of type 2N 3702 (*pnp*) or 2N 3704 (*npn*) is suitable each with a power limit of 300 mW at 25°C in free air.

Collector load. In making use of eqn. (4.10), let V_{CC} be reduced by 0.5 V and consider a total power output of 1.5 W,

$$R_C = (14.5)^2/(2 \times 1.5) = 210.25/3 = 70.08 \ \Omega.$$

Each transistor sees half the full load for half the time. Hence, the collector-to-collector load is $4 \times 70.08 = 280.3 \ \Omega$. The required transformer turns ratio to match a 3-Ω speaker to this load is $\sqrt{280/3} = 9.6:1$.

Biasing. Since the emitter resistor reduces efficiency its value is kept low; let it be the preferred value of 4.7 Ω. Checking the output, the total load on each transistor is

$$R_C + R_E = 70.08 + 4.7 = 74.78 \ \Omega.$$

Total power $P_{o(max)} = V_{CC}^2/2(R_C + R_E) = 210.25/149.56 \doteqdot 1.4$ W. Useful power output $= P_{o(max)} \times R_C/(R_C + R_E) = 1.4 \times 0.93 = 1.3$ W, which meets the specification. The values of R_1 and R_2 are chosen bearing in mind the required bias voltage, let this be 150 mV, and the amount of current drain which may be permitted. Limiting the current drain to 1.5 mA, which is comparable to the quiescent current of the transistors, then,

$$R_1 + R_2 = V_{CC}/1.5 \ \text{mA} = 10 \ \text{k}\Omega.$$

For $V_B = 150$ mV, $R_1 = 99R_2$. As this is not critical, and bearing in mind the spreads likely to be met in the transistors, let $R_1 = 9.1\,k\Omega$ and $R_2 = 100\,\Omega$.

For increased stability R_2 may be changed to a 220-Ω resistor shunted by an N.T.C. thermistor having a nominal resistance of 200 Ω.

Summing up the final design; turns ratio of output transformer 4.8 + 4.8 to 1, $R_E = 4.7\,\Omega$, $R_1 = 9.1\,k\Omega$ and $R_2 = 220\,\Omega$ shunted by a negative temperature coefficient thermistor of 200 Ω.

4.4. The capacitively coupled amplifier

In applications where an amplifier is not required to operate on very low frequencies (below 1 Hz) capacitor coupling can be used. This isolates the d.c. voltage of a collector from the base of the following stage.

Bandwidth

The capacitively coupled amplifier has a finite bandwidth (see Fig. 4.7). The gain falls off as the reactance of the coupling capacitor

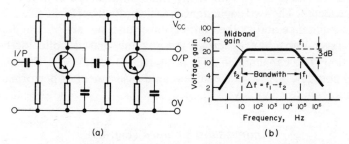

Fig. 4.7. (a) A capacitively coupled amplifier stage. (b) Gain plotted against frequency. The upper and lower bandwidth limits are f_1 and f_2, i.e. the frequencies at which the gain has fallen to $1/\sqrt{2}$ of the midband value.

increases at low frequency, and also at high frequency due to the shunting effect of transistor capacitances. The bandwidth is defined as the frequency range between the two frequencies for which the gain is 3 dB down on the midband gain.

4.5. High-frequency performance

The transistor can be represented by the hybrid π equivalent network of Fig. 4.8a, in which $r_{bb'}$ is the ohmic resistance between the active base region and the base lead. To investigate its high-frequency performance the network can be modified to form

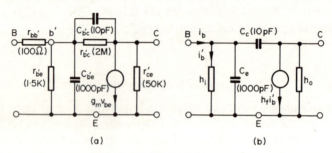

(a)   (b)

Fig. 4.8. Modified hybrid network.

another, similar to the low-frequency h parameter network with the feedback generator zero. This is a reasonable approximation as the input resistance does not change greatly, for even quite large load resistance variations, as indicated in Fig. 1.37.

The resistance $r_{bb'}$ is sufficiently small, in this type of network, to be neglected. The two capacitances which affect performance are c_e, the capacitance across the emitter–base junction, and c_c the depletion layer capacitance across the reverse biased collector–base junction. The capacitance c_c is proportional to $1/\sqrt{V_{CB}}$. The simplified network of Fig. 4.8b enables rapid assessment of performance with adequate accuracy.

Output short circuited (Fig. 4.9)

The feedback capacitance c_c can be neglected as it is shunted by c_e which is 100 times larger. At low frequencies $i'_b = i_b$, but at higher frequencies h_i is bypassed by c_e and the output current $h_f i'_b$ is reduced,

$$i'_b = \frac{1/sc_e}{h_i + (1/sc_e)} i_b.$$

Output current $i_c = h_f i'_b = \dfrac{h_f i_b}{1 + sc_e h_i},$ (4.12)

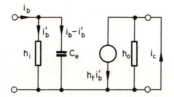

Fig. 4.9. Equivalent network with output short circuited.

$$i_c/i_b = h_f/(1 + s\tau_B), \quad \text{where } \tau_B = c_e h_i.$$

and the short-circuit current gain,

$$\frac{i_c}{i_b} = \frac{h_f}{1 + s\tau_B} \quad \text{(where } \tau_B = c_e h_i). \tag{4.13}$$

Thus, if $h_i = 2.2\,\text{k}\Omega$ and $c_e = 700\,\text{pF}$, the time constant $\tau_B = 1.54 \times 10^{-6}\,\text{sec}$. If a *breakpoint* is defined as $\omega_B = 1/\tau_B$, then the frequency of the breakpoint is

$$f_B = \frac{10^6}{2\pi \times 1.54}\,\text{kHz} = 103\,\text{kHz}.$$

It will be seen later (§ 4.7) that the relative short-circuit current gain can be represented by a simple lag with this breakpoint, as shown in the frequency response curve of Fig. 4.10.

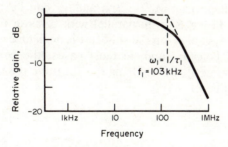

Fig. 4.10. Relative gain against frequency for short-circuit output. Input is considered as a passive lag.

Load conductance G_L

When the output is not a short circuit, as in Fig. 4.11, the effect of c_c may not be negligible.

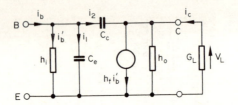

Fig. 4.11. Network including the load G_L.

The current in c_c is now

$$i_2 = (v_b - v_c)sc_c \quad \text{where} \quad v_c = -i_c/G_L. \tag{4.14}$$

But,

$$i_b' = \frac{v_b}{h_i} \quad \text{and} \quad v_c = \frac{-h_f}{h_o + G_L} \cdot \frac{v_b}{h_i},$$

or

$$v_c = -g_m' R v_b \left(\text{where} \quad g_m' = \frac{h_f}{h_i} \quad \text{and} \quad R = \frac{1}{h_o + G_L} \right).$$

Substituting for v_c in eqn. (4.14),

$$i_2 = (1 + g_m R)sc_c v_b = (1 - A_v)sc_c v_b,$$

where the voltage gain $A_v = v_c/v_b$, and c_c can be replaced by a capacitor $c_b = (1 + g_m R)c_c$ in parallel with c_e as in Fig. 4.12a.

The direct effect of c_c on the collector is negligible compared with the base time constant and, in most cases, the collector can be represented as in Fig. 4.12b. Using this diagram, the h.f. breakpoint can be determined.

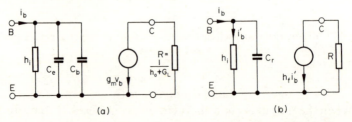

(a)   (b)

Fig. 4.12. Simplified representation of a transistor with a resistive load.

Current gain at h.f.

$$A_i = \frac{A_o}{1 + s\tau_1},\tag{4.15}$$

where

$$A_o = \frac{G_L}{h_o + G_L} \quad \text{and} \quad \tau_1 = h_i C_T.$$

Example. $h_i = 2.2\,\text{k}\Omega$, $c_e = 700\,\text{pF}$, $c_c = 10\,\text{pF}$, $R_C = 1\,\text{k}\Omega$, $h_f = 100$ and $1/h_o = 25\,\text{k}\Omega$.

$$A_v = -\frac{h_f}{h_i(G_L + h_o)} = -\frac{100}{2.2 \times 10^3 \times (1000 + 40)10^{-6}},$$

$$= -44.$$

$$c_b = (1 - A_v)c_c = 45 \times 10\,\text{pF} = 450\,\text{pF}.$$

$$C_T = c_e + c_b = 1150\,\text{pF}.$$

$$\tau_1 = h_i C_T = 2.2 \times 10^3 \times 1150 \times 10^{-12}$$

$$= 2.5 \times 10^{-6}\,\text{sec}.$$

$$\omega_1 = 1/\tau_1 = 400 \times 10^3\,\text{r/s}.$$

$$f_1 = \omega_1/2\pi = 64\,\text{kHz}.$$

Thus, if h_i, h_f, h_o, c_e and c_c are known the high-frequency breakpoint can be estimated.

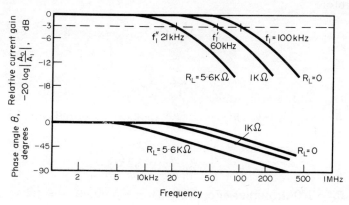

Fig. 4.13. Frequency response for a typical alloy transistor with various collector resistors.

The frequency response curves for a typical alloy transistor with various collector resistors are shown in Fig. 4.13.

Measurement of c_e and c_c

Values of c_e and c_c for a particular transistor can be found by finding the time constant τ_1,

(a) with output short circuit $\tau_1(a)$,
(b) with known load resistance $\tau_1(b)$.

The time constant τ_1 is determined by finding the frequency at which the current gain A_o falls by 3 dB to $0.7 A_o$ in each of the above cases. For case (a),

$$h_i = \frac{v_{be}}{i_b} \quad \text{and} \quad c_e = \frac{\tau_{1(a)}}{h_i}. \tag{4.16}$$

Time constant $\tau_{1(b)}$ is found for a load resistance R_C giving a voltage gain of the order of 100 ($A_v = v_{ce}/v_{be}$).

$$C_T = \tau_{1(b)}/h_i, \tag{4.17}$$

$$c_b = C_T - c_e. \tag{4.18}$$

But,

$$C_T = (1 - A_v)c_c + c_e. \tag{4.19}$$

Therefore

$$c_b = (1 - A_v)c_c, \tag{4.20}$$

or

$$c_c = \frac{C_T - c_e}{1 - A_v}. \tag{4.21}$$

Example. $A_v = (v_{ce})/(v_{be}) = -209$ (negative sign indicates signal inversion).

$$h_i = \frac{v_{be}}{i_b} = \frac{22 \text{ mV}}{10 \text{ } \mu\text{A}} = 2.2 \text{ k}\Omega.$$

For $R_C = 100 \text{ }\Omega$ (output effectively short circuited), $f_1 = 100$ kHz.

$$\omega_1 = 628 \times 10^3 \text{ r/s}$$

and

$$\tau_1 = 1.6 \times 10^{-6} \text{ sec.}$$

Therefore,

$$c_e = \frac{1.6 \times 10^{-6}}{2.2 \times 10^3} = 730 \text{ pF} \quad \text{[using eqn. (4.16)].}$$

For $R_C = 5.6 \text{ k}\Omega$, $f_1 = 21 \text{ kHz}$, $\omega_1 = 132 \times 10^3$, and $\tau_1 = 7.6 \times 10^{-6}$ sec. Therefore,

$$C_T = \frac{7.6 \times 10^{-6}}{2.2 \times 10^{-3}} = 3460 \text{ pF} \quad \text{[using eqn. (4.17)].}$$

$$c_b = C_T - c_e = 2730 \text{ pF} \quad \text{[eqn. (4.18)].}$$

$$c_c = \frac{c_b}{1 - A_v} = \frac{2730}{210} = 13 \text{ pF} \quad \text{[from eqn. (4.20)].}$$

4.6. High-frequency response

The name *transfer function* is a term which describes the complex ratio of the output signal of a device (or system), to its input signal, i.e. these quantities are stated in terms of the Laplace operator $s (= \alpha + j\omega)$. Thus, eqn. (4.15) has the form of a transfer function. In such an equation $j\omega$ may be substituted for s, if only sinusoidal signals are being considered. The frequency response may be obtained in this way. Thus, from eqn. (4.15),

$$A_1(j\omega) = A_0 \left(\frac{1}{1 + j\omega\tau_1}\right),$$

and rationalizing,

$$A_1(j\omega) = A_0 \left(\frac{1}{1 + (\omega\tau_1)^2}\right)(1 - j\omega\tau_1)$$

$$= |A_1(j\omega)| \angle \theta_1,$$

where,

$$|A_1(j\omega)| = \frac{A_0}{\sqrt{[1 + (\omega\tau_1)^2]}}, \tag{4.22}$$

and

$$\theta_1 = -\tan^{-1} \omega\tau_1.$$

At very high frequencies, i.e. frequencies at which $\omega \gg 1/\tau_1$ (and consequently $\omega\tau_1 \gg 1$), the real part of the gain expression is very small and $A_1(j\omega)$ tends to $-j(A_0/\omega\tau_1)$. This implies that the phase is retarded 90° on the midband value, and gain falls inversely with frequency. The locus of the gain lies on the circumference of a semicircle with midband gain as diameter (see Fig. 4.14).

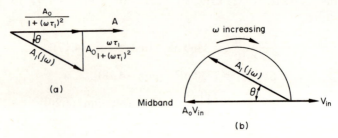

(a)

(b)

Fig. 4.14. (a) Resolved components of high frequency gain. (b) Polar diagram of output voltage as a function of ω. As ω becomes very large θ tends to 90°.

Bandwidth limit

The upper bandwidth limit is the frequency at which the gain falls to $1/\sqrt{2}$ of the midband value. From eqn. (4.22), the upper bandwidth frequency $\omega_1 = 1/\tau_1$, since

$$|A_1(j\omega)| = \frac{A_0}{\sqrt{1+1}} = \frac{A_0}{\sqrt{2}}.$$

The angle of the phasor relative to A_0,

$$\theta_1 = -\tan^{-1}\omega\tau_1 = -\tan^{-1} 1 = -45° \text{ (see Figs. 4.10 and 4.15).}$$

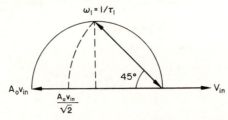

Fig. 4.15. Phasor for $\omega_1 = 1/\tau_1$; ω_1 is the radian frequency at which A falls to $A_0/\sqrt{2}$.

4.7. Asymptotic approximation

The frequency response of a system can be estimated by the use of straight-line asymptotes. Let the midband gain be A_0 and the high-frequency gain be A_1. Then, from eqn. (4.22) the magnitude of gain

$$|A_1(j\omega)| = \frac{A_0}{\sqrt{[1+(\omega\tau_1)^2]}}.$$

At frequencies where $\omega\tau_1$ is very much greater than unity

$$|A_1(j\omega)| \doteq \frac{A_0}{\omega\tau_1}.$$

Gain falls as ω increases. If the magnitude of the gain at frequency ω_a is A_a then at frequency $2\omega_a$ the gain will be $A_a/2$.

Doubling the frequency is an *octave* change and thus the gain is halved for each octave. By similar reasoning, a decade increase in frequency reduces the gain by one tenth. As shown in Fig. 4.16, it is usual to employ logarithmic axes so that at frequencies $\omega \gg 1/\tau_1$ the gain curve is a straight line:

$$20 \log_{10} \left| \frac{A_1(j\omega_a)}{A_2(j2\omega_a)} \right| = 20 \log_{10} \frac{A_a}{A_a/2}$$
$$= 20 \log_{10} 2 = 6 \text{ dB}.$$

That is, the gain falls at a rate of 6 dB/octave, or 20 dB/decade. The straight line representing $|A_1(j\omega)| \doteq A_0/\omega\tau_1$ is shown in Fig. 4.16. If the line is projected back so that it meets the midband gain line, it will do so at $\omega_1 = 1/\tau_1$. In § 4.5 it was shown that at ω_1 the gain had fallen from A_0 to $A_0/\sqrt{2}$, a fall of 3 dB, since $20 \log_{10} \sqrt{2} = 3$. Thus, the error in taking the straight-line approximation of the amplitude response is, as shown in Fig. 4.16, to be a maximum of 3 dB at ω_1 falling to 1 dB at $\omega_1/2$ and $2\omega_1$.

It is convenient to draw the amplitude response relative to the gain A_0. A negative sign shows that the gain is less than A_0 and the bandwidth is given by the frequencies at which the relative gain is -3 dB:

$$\text{relative gain} = -20 \log_{10} \left| \frac{A_0}{A(j\omega)} \right|.$$

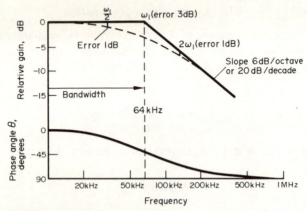

Fig. 4.16. Frequency response using straight-line asymptotes. Gain is relative to midband. Straight-line approximation is made by projecting down from ω_1 ($= f_1/2\pi$) on the 0 dB axis, at a slope of 20 dB/decade. Note that the ordinate is gain relative to A_0.

The straight-line approximation for $A = |A_0/(1 + j\omega\tau_1)|$ can be drawn by projecting down from ω_1 on the zero dB relative gain line at a slope of 6 dB for each octave of frequency. Often it is more convenient to use the slope of 20 dB/decade, as the relative gain axis is usually calibrated in tens and the logarithmic frequency scales repeat in cycles of ten.

Thus, referring to eqn. (4.15) and its subsequent example, the time constant τ_1 was found to be 2.5×10^{-6} s. From this, the breakpoint $\omega_1 = 1/\tau_1 = 400 \times 10^3$ r/s, and the frequency is $\omega_1/2\pi = 64$ kHz. The asymptotic approximation of the frequency response is therefore a straight line at 0 dB up to the breakpoint, followed by a slope falling at 6 dB/octave or 20 dB/decade.

4.8.　Low-frequency performance of capacitively coupled stages

As the frequency is reduced, the increased impedance of C_c reduces i_{in}, the transistor input current (see Fig. 4.17). At frequencies where r_{in} is much greater than $1/\omega C_c$,

$$i_{in} = \frac{R_{out}}{R_{out} + r_{in}} \cdot i \quad \text{(if } R_B \gg r_{in}\text{)},$$

and for efficient current transfer, r_{in} should be much less than R_{out}.

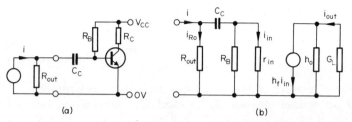

Fig. 4.17. Capacitively coupled amplifier with low-frequency equivalent network.

If i is the current from the transistor of the previous stage, R_{out} will be the resistance of $1/h_o$ in parallel with R_C for that stage. If these two are both $20\,\text{k}\Omega$, R_{out} will be $10\,\text{k}\Omega$ and, for r_{in} of $2\,\text{k}\Omega$ the current transfer $i_{in}/i = 10\,\text{k}\Omega/(10\,\text{k}\Omega + 2\,\text{k}\Omega) = 5/6$.

For efficient transfer, R_C should be as large as possible. At low frequencies

$$i_{in}(s) = \frac{R_{out}}{R_{out} + r_{in} + (1/sC_c)} \cdot i(s). \tag{4.23}$$

The current transfer falls by 3 dB when $R_{out} + r_{in} = 1/\omega C_c$ or

$$\omega = \frac{1}{C_c(R_{out} + r_{in})} = \frac{1}{\tau_2}, \tag{4.24}$$

where $\tau_2 = C_c(R_{out} + r_{in})$.

Equation (4.23) can be rewritten with the frequency invariant components extracted:

$$i_{in}(s) = \frac{R_{out}}{R_{out} + r_{in}} \cdot \frac{1}{1 + (1/s\tau_2)} i(s). \tag{4.25}$$

$$i_{out}(s) = A_0 i_{in} \quad \left(\text{where } A_0 = \frac{h_f G_L}{h_o + G_L}\right),$$

$$= A_0 \frac{R_{out}}{R_{out} + r_{in}} \cdot \frac{1}{1 + (1/s\tau_2)} i(s),$$

$$= A_2 \frac{1}{1 + (1/s\tau_2)} i(s),$$

$$\left(\text{where } A_2 = A_0 \cdot \frac{R_{out}}{R_{out} + r_{in}}\right). \tag{4.26}$$

For coupling capacitor C_c,

$$A(s) = A_2 \frac{1}{1 + (1/s\tau_2)}. \tag{4.27}$$

Effect of emitter capacitor on low-frequency performance

If an emitter resistor R_E is used for stabilization (see § 1.15), usually a bypass capacitor C_E will be employed to increase gain at signal frequencies.

At sufficiently low frequency (or if C_E is not present), the input resistance of the transistor is increased from h_i to $h_i + (1 + h_f)R_E$. This follows from considerations given in § 1.16 and shown in Fig. 4.18:

$$z_{in}(s) = h_i + (1 + h_f)Z_E,$$

$$= h_i + (1 + h_f)\frac{R_E}{1 + sC_E R_E}.$$

This should be inserted in place of r_{in} for the expression for A_2 in eqn. (4.26).

Thus,

$$A_2(s) = \frac{R_{out}}{R_{out} + z_{in}} A_0,$$

$$= \frac{R_{out}}{R_{out} + h_i + (1 + h_f)[R_E/(1 + sC_E R_E)]} A_0,$$

$$= \frac{R_{out}}{R_{out} + h_i + (1 + h_f)R_E}$$

$$\times \frac{1 + sC_E R_E}{1 + \{sC_E R_E(R_{out} + h_i)/[R_{out} + h_i + (1 + h_f)R_E]\}} A_0. \tag{4.28}$$

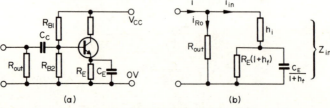

(a)   (b)

Fig. 4.18. Input network neglecting bias resistors and coupling capacitor. R_0 includes R_B.

Thus, the gain expression for partially bypassed emitter resistor is

$$A(s) = A_E \frac{1 + s\tau_E}{1 + s\tau_E'},$$

where

$$A_E = \frac{R_{out}}{R_{out} + h_i + (1 + h_f)R_E} A_0, \qquad (4.29)$$

$$\tau_E = C_E R_E \quad \text{and} \quad \tau_E' = \frac{\tau_E(R_{out} + h_i)}{R_{out} + h_i + (1 + h_f)R_E},$$

$$\tau_E > \tau_E'.$$

This is illustrated in Fig. 4.19.

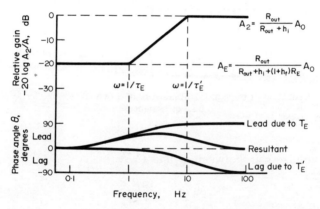

Fig. 4.19. Effect of emitter time constant on amplifier performance. $A_2/A_E = 10$.

4.9. Tandem stages

If one transistor is followed by another, the first has the input impedance of the second as its load impedance.

In Fig. 4.20b, the effect of the collector capacitance c_{c2} is to augment c_{e2} giving rise to C_T of Fig. 4.20c.

Part of the input current will flow in c_{c1}, the amount being determined by the voltage gain between the base and collector. The voltage gain of T_1 is complex, because the collector load is G and C_T in parallel.

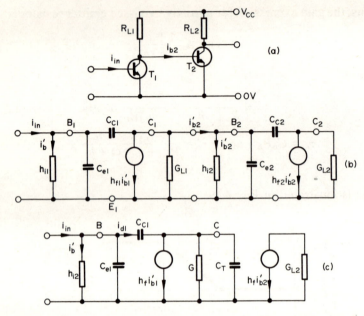

Fig. 4.20. Reduction of the equivalent network of a two-stage amplifier to estimate h.f. performance.

The voltage gain between B and C,

$$A_{v1} = -\frac{g_m R}{1 + s\tau},$$ (4.30)

where $\tau = C_T/G$ and $R = 1/G$.

Thus,

$$i_{d1} = v_{be}\left(1 + \frac{g_m R}{1 + s\tau}\right) s c_{c1},$$

and the impedance to be shunted across the base of T_1 by c_{c1} is

$$Z(s) = \frac{1 + s\tau}{[(1 + g_m R) + s\tau] s c_{c1}},$$

or,

$$Z(j\omega) = \frac{1 + j\omega\tau}{j\omega c_{c1}(1 + g_m R + j\omega\tau)}.$$ (4.31)

Separating real and imaginary components,

$$Z(j\omega) = \frac{1 + g_m R + (\omega\tau)^2}{j\omega c_{c1}(1 + g_m R)^2 + (\omega\tau)^2}$$
$$+ \frac{\tau g_m R}{c_{c1}[(1 + g_m R)^2 + (\omega\tau)^2]}, \qquad (4.32)$$
$$\doteqdot \frac{1}{j\omega c_{c1} \cdot g_m R} + \frac{C_T}{c_{c1} g_m} \quad \text{as } g_m R \gg 1, \qquad (4.33)$$

and for frequencies where $\omega\tau < 1$.

The added impedance is a resistance $C_T/c_{c1}g_m$ in series with capacitance $c_{c1}g_m R$. The resulting simplified network is shown in Fig. 4.21.

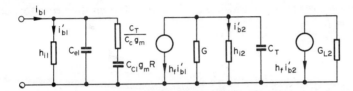

Fig. 4.21. Equivalent network of a two-stage amplifier with collector-to-base coupling eliminated. Note that a resistance and capacitance in series appears across the input.

DESIGN EXAMPLE 4.4

Required, an amplifier stage with transfer resistance of $100 \text{ k}\Omega$ and bandwidth from 50 Hz to 20 kHz. The signal source resistance is $5 \text{ k}\Omega$ and a peak output of 4 V is required. (Transfer resistance $= v_{out}/i_{in}$).

Using emitter stabilization, a starting-point is $R_B/R_E = 5$ (§ 1.15) where

$$R_B = \frac{R_{B1} R_{B2}}{R_{B1} + R_{B2}}.$$

For a nominal $h_f = 50$ (minimum value for a 2N 1304),

$$K = 1 + \frac{h_f R_E}{R_E + R_B} = 1 + \frac{50 \times 1}{1 + 5} = 9.3.$$

Supply voltage. To obtain an output voltage swing of 8 V peak to peak, $V_{CC} = 12$ V is suitable.

Collector resistor.

$$R_T = \frac{h_f}{G_C + h_o} \div \frac{h_f}{G_C} \quad \text{(for } h_o \ll G_C \text{)}.$$

The transfer resistance of $100 \, \text{k}\Omega$ requires a collector resistor of $2.2 \, \text{k}\Omega$. Let the value $2.7 \, \text{k}\Omega$ be used to allow for the reduction in current gain by the bias network.

Collector current. For the required output voltage, make $V_C = 7.5 \, \text{V}$,

$$I_C = \frac{V_{CC} - V_C}{R_C} = \frac{4.5 \, \text{V}}{2.7 \, \text{k}\Omega} = 1.7 \, \text{mA}.$$

Emitter resistor. Allowing $2.5 \, \text{V}$ across the emitter resistor, and as $I_E = I_C$,

$$R_E \div \frac{V_E}{I_E} = \frac{2.5 \, \text{V}}{1.7 \, \text{mA}} = 1.5 \, \text{k}\Omega.$$

Base resistors. R_B, which is R_{B1} and R_{B2} in parallel $= 5R_E = 7.5 \, \text{k}\Omega$,

$$V_B = V_{BE} + V_E,$$
$$= 0.15 \, \text{V} + 2.5 \, \text{V} \quad \text{for a germanium transistor,}$$
$$= 2.65 \, \text{V}.$$

Neglecting the base current I_B,

$$\frac{V_B}{V_{CC}} = \frac{2.65 \, \text{V}}{12 \, \text{V}} = 0.22,$$

$$= \frac{R_{B1}}{R_{B1} + R_{B2}} = \frac{R_B}{R_{B2}} = \frac{5R_E}{R_{B2}}.$$

Thus,

$$R_{B2} = \frac{5R_E}{0.22} = \frac{5}{0.22} \times 1.5 \, \text{k}\Omega = 34 \, \text{k}\Omega.$$

Also,

$$\frac{R_{B1}}{R_{B2}} = \frac{2.65 \, \text{V}}{9.35 \, \text{V}},$$

therefore,

$$R_{B1} = \frac{2.65}{9.35} \times 34 \text{ k}\Omega = 9.65 \text{ k}\Omega.$$

Let $R_{B1} = 9.1$ kΩ and $R_{B2} = 33$ kΩ. The d.c. conditions are shown in Fig. 4.22a. R_B shunts the input of the transistor and reduces the gain, as in Fig. 4.22b.

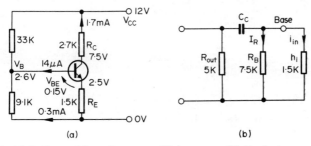

(a)   (b)

Fig. 4.22. (a) D.C. conditions for an amplifying stage. (b) Equivalent network to show the effect of R_B shunting h_i, assuming R_E to be fully bypassed by C_E.

The input resistance h_i to the transistor is nominally 1.5 kΩ for a collector current of 1.7 mA.

$$R'_T = \frac{R_E}{R_B + h_i} \cdot \frac{h_f}{G_C + h_o},$$

$$= \frac{7.5 \text{ k}\Omega}{7.5 \text{ k}\Omega + 1.5 \text{ k}\Omega}$$

$$\times \frac{50}{370 \times 10^{-6} + 50 \times 10^{-6}} \doteqdot 100 \text{ k}\Omega.$$

Coupling capacitor

$\tau_2 = C_c(R_{out} + r_{in}) = 1/\omega_2$ from eqn. (4.24), where r_{in} is resistance R_B and h_i in parallel (= 1.1 kΩ). It is required that $f_2 = 50$ Hz, or $\omega_2 = 314$ r/s.

Therefore

$$\tau_2 = 3.19 \times 10^{-3} \text{ s}.$$

$$C_c = \frac{\tau_2}{R_{out} + r_{in}} = \frac{3.19 \times 10^{-3}}{5 \text{ k}\Omega + 1.1 \text{ k}\Omega} = 0.523 \ \mu\text{F}.$$

Let C_C be a 1-μF electrolytic capacitor.

Emitter resistor bypass capacitor

$$\tau'_E = \frac{\tau_E(R'_{out} + h_i)}{R'_{out} + h_i + (1 + h_f)R_E}, \quad \text{from eqn. (4.29)}$$

where R'_{out} is R_{out} in parallel with R_B, as in Fig. 4.23.

$$\frac{\tau'_E}{\tau_E} = \frac{R'_{out} + h_i}{R'_{out} + h_i + (1 + h_f)R_E},$$

$$= \frac{3 \,\text{k}\Omega + 1.5 \,\text{k}\Omega}{3 \,\text{k}\Omega + 1.5 \,\text{k}\Omega + (1 + 50)1.5 \,\text{k}\Omega},$$

$$\doteqdot 0.06.$$

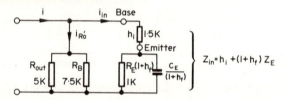

Fig. 4.23. Division of input current between R'_{out} and base.

To ensure that the emitter impedance does not affect the gain at 50 Hz let τ'_E be $5\tau_2 = 1.6 \times 10^{-2}$ sec. The breakpoint $1/\tau'_E$ will be at 62.8 r/s, i.e. 10 Hz.

$$\tau'_E = \tau_E \times 0.06,$$

therefore

$$C_E = \frac{\tau_E}{R_E} = \frac{1.6 \times 10^{-2}}{0.06 \times 1.5 \times 10^3} = 175 \,\mu\text{F}.$$

Let C_E be 200 μF.

High-frequency performance. This is a function of the transistor capacitances c_e and c_c.

From Fig. 4.13 it is apparent that, with $R_C = 2.7 \,\text{k}\Omega$, a 20 kHz upper frequency limit can be obtained. The completed design is shown in Fig. 4.24.

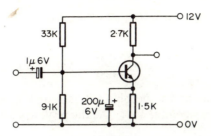

Fig. 4.24. Capacitively coupled amplifier stage of Design Example 4.4.

DESIGN EXAMPLE 4.5

Required, an amplifier with voltage gain greater than $-50,000$ over a frequency range of 100 Hz to 10 kHz. The required peak output voltage is 10 V.

Voltage gain. $A_v = R_L A_i / r_{in}$, where A_i is the over-all current gain, R_L is the effective load resistance, and r_{in} is the input resistance to the amplifier.

To provide the necessary current gain and signal inversion, three stages are required.

Let the transistors be 2S 103 which are silicon mesa types with voltage rating $V_{CEO} = 40$ V and $h_{fe(min)} = 50$. As shown in Fig. 4.25, these can be directly coupled (see § 4.12). Estimating the minimum effective current gain of each stage as 30, the overall current gain $A_i = 27,000$.

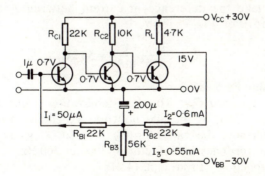

Fig. 4.25. Direct coupled three stage amplifier with zero frequency stabilization.

Load resistor R_L. If the input transistor is run at a current that gives $r_{in} = 2\,\text{k}\Omega$,

$$R_L = \frac{A_v}{A_i}\,r_{in} \doteq \frac{50,000}{27,000} \times 2\,\text{k}\Omega = 3.7\,\text{k}\Omega.$$

To allow for loading by the bias network and h_o, let $R_L = 4.7\,\text{k}\Omega$.

Collector supply voltage V_{CC}. As the peak to peak output voltage is 20 V let $V_{CC} = 30\,\text{V}$.

Collector resistors. For T_1 to be run at 1.5 mA, $R_{c1} = 30\,\text{V}/1.5\,\text{mA} = 20\,\text{k}\Omega$. For T_2 to be run at 3.0 mA, $R_{c2} = 30\,\text{V}/3.0\,\text{mA} = 10\,\text{k}\Omega$.

Bias network. At operating frequency, R_{B1} and R_{B2} shunt the input and output respectively. To minimize shunting effects let them both be 22 kΩ, as in Fig. 4.25.

The estimated input current

$$I_1 = \frac{I_{c1}}{h_{fe}} \doteq \frac{1.5\,\text{mA}}{30} = 50\,\mu\text{A}.$$

This current, flowing in R_{B1}, will produce 1.1 V, which must be added to $V_{BE} = 0.7\,\text{V}$ to give the voltage at the junction of R_{B1} and R_{B2}. This is approximately 2 V.

For a collector voltage for T_3 of 15 V, the current in R_{B2},

$$I_2 = \frac{15\,\text{V} - 2\,\text{V}}{22\,\text{k}\Omega} \doteq 0.6\,\text{mA}.$$

R_{B3} must take the difference in current between I_2 and I_1, i.e. $I_3 \doteq 0.55\,\text{mA}$. For $V_{BB} = -30\,\text{V}$,

$$R_{B3} = \frac{30\,\text{V} + 2\,\text{V}}{0.55\,\text{mA}} = 58\,\text{k}\Omega.$$

Let the value be 56 kΩ.

Bypass capacitor C_B. The d.c. stabilization is due to negative feedback provided by the bias network. To eliminate feedback at signal frequency, the capacitor C_B is used. A value of 200 μF reduces the feedback by a factor of 3000 at 100 Hz.

Input capacitor. From eqn. (4.24),

$$\omega = \frac{1}{C_C(R_{out} + r_{in})},$$

or $C_C = 1/\omega r_{in}$ if R_{out}, the source resistance, is zero. (This gives the maximum value of C_C.)

For $\omega = 628\,r/s$ $(f = 100\,Hz)$,

$$C_C = \frac{1}{628 \times 2\,k\Omega} = 0.8\,\mu F.$$

$$C_C = 1\,\mu F.$$

The completed circuit is shown in Fig. 4.25.

4.10. Amplifier time response[36]

Complementary to the frequency response of a network is its behaviour to a step input. This is the *time response*, as the amplitude of the output signal varies with time.

The expressions obtained as functions of the complex variable s can be transformed into functions of time as follows.

Effect of shunt capacitance (see Fig. 4.12)

When a step voltage is applied to the input of an amplifier, the output can rise only as C_T is charged. Thus, for a fast rise, the time constant $C_T h_i$ must be as small as possible.

$$v_{out}(s) = \frac{A_0}{(1 + s\tau_1)}\,v_{in}(s) \quad \text{[from eqn. (4.15)],}$$

$$= \frac{A_0}{\tau_1} \cdot \frac{1}{s(s + 1/\tau_1)} \quad \text{(for unit step input).}$$

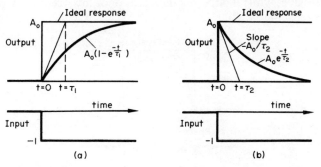

Fig. 4.26. Response to a negative step input showing the effect on performance of (a) shunt capacitance and (b) coupling capacitance.

Thus,

$$v_{out}(t) = A_0[1 - \exp(-t/\tau_1)] \quad \text{(from transform pair No. 3)}$$

$$(4.34)$$

This is plotted in Fig. 4.26a.

Effect of coupling capacitance (see Fig. 4.17)

When a step function is applied to the coupling capacitor the full value appears at the base terminal. As the capacitor charges, this voltage falls to zero on a time constant $\tau_2 = C_c(R_{out} + r_{in})$ as shown in Fig. 4.26b:

$$v_{out}(s) = \frac{A_0}{1 + (1/s\tau_2)} \, v_{in}(s).$$

For a negative step function input,

$$v_{out}(t) = A_0 \exp(-t/\tau_2) \quad (4.35)$$

(from transform pair No. 2).

In most applications τ_2 will be much greater than the length of any pulse τ that constitutes the input signal. The top of the pulse falls linearly with time and the slope is proportional to $1/\tau_2$. If C_C is infinite (or the amplifier is direct coupled) the slope will be zero. The fall in output amplitude is called "sag".

Combined effects of coupling and shunt capacitance

If the high-frequency performance is inadequate, the rise time will be excessive, while a large sag indicates poor low-frequency performance (see Fig. 4.27).

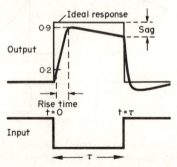

Fig. 4.27. Pulse response showing the effect of C_s and C_c.

4.11.   Zero frequency amplifiers

It is often required, particularly in control-engineering problems, that an amplifier should amplify down to zero frequency. For a capacitor coupled amplifier this would require the use of an infinite capacitor. A zero frequency amplifier is able to amplify a change in d.c. level, as shown in Fig. 4.28.

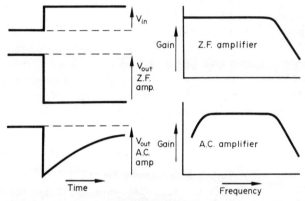

Fig. 4.28. Comparison of time and frequency responses of zero frequency and a.c. amplifiers.

There are two general types of z.f. amplifiers:

(a) *Direct coupled.*   One stage is coupled to the next, either directly or by means of a resistance chain. The arrangement of Fig. 4.25 is an example of the former.

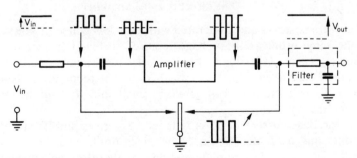

Fig. 4.29. Schematic diagram to illustrate the operation of a chopper z.f. amplifier. Although a mechanical relay driven by an a.c. source is indicated here, chopping would be achieved by the use of a solid state switching device.

(b) *Modulator* (*chopper*).   The input is chopped, amplified as an a.c. signal and reconstituted at the output, as shown in Fig. 4.29.

Type (a) will usually have a greater bandwidth than type (b), but the latter has less output drift.

Drift

With no input signal, the output from a z.f. amplifier can alter due to changes of operating conditions. These changes are amplified in the same way as input signals and cannot be separated from them. The change in output voltage for zero input is known as *drift*. Drift is measured by shorting the input terminals and monitoring the output, and is typically given in volts per hour; so that different amplifiers may be compared, drift is usually referred to the input.

$$\text{Input drift } (V_{DI}) = \frac{\text{Output drift } (V_{DO})}{\text{amplifier gain (A)}}. \tag{4.36}$$

Drift signals are of very low frequency having a period which may vary from several seconds to hours, and are consequently not passed by a.c. amplifiers.

Zeroing

Since the output level is of significance, with a z.f. amplifier, some means is required of setting the output to the required datum when the input is zero. (For example, see Fig. 4.35.)

4.12.   The direct coupled amplifier

Most transistors can be operated with very low collector voltages, of the same order as the voltage required to bias the base of the following stage. Referring to Fig. 4.30, if V_{CC} is much greater than V_{CE}, then the current $I = (V_{CC} - V_{CE})/R_C$ will be relatively constant. When T_1 is bottomed V_{BE} is very small and T_2 will be non-conducting.

As the base current of T_1 is reduced, V_{BE} rises and T_2 starts to conduct and as I_1 decreases, I_2 increases until, when T_1 is non-conducting, $I_2 = I$. The amplifier action may therefore be considered as the exchange of I between I_1 and I_2 due to the variations in T_1

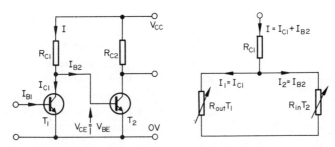

Fig. 4.30. Direct coupled amplifier with network to represent the division of current between the collector of T_1 and the base of T_2.

output resistance R_{out}, and R_{in}, the input resistance of T_2. R_{in} is the forward-biased diode resistance of the base–emitter junction and, due to its characteristic, the voltage V_{BE} across it is limited. Since R_C is made much greater than R_{in}, R_{in} is effectively the load resistance of T_1.

Example. Let $R_{C1} = 15\,k\Omega$, $R_{C2} = 3.9\,k\Omega$, $V_{CC} = 6\,V$ and $I = 0.4\,mA$. The load line may be constructed as in Fig. 4.31a, in the following manner. The input characteristic *OAB* of T_2 is superimposed on the collector characteristic of T_1, and then rotated about the axis drawn at $I_C = I/2$ (in this case 0.2 mA), giving the curve *CAD*. The line *CB* represents the constant current $I = 0.4\,mA$.

For any voltage V_C, a vertical line joining the voltage axis to the

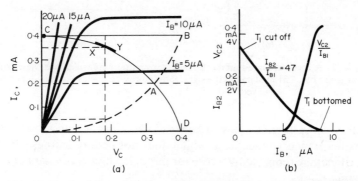

Fig. 4.31. (a) The load line of T_1 (from Fig. 4.30), is the curve *CD* of which the section *XY* is the region of operation. From this load line the current gain for the first stage can be derived. Also shown (b) is the transfer characteristic V_{C2}/I_{B1}.

line CB is divided into two parts such that the part of the vertical above the curve CAD represents the current I_{B2}, the part below being I_{C1}. Thus, at $V_C = 0.18$ V, $I_{C1} = 0.35$ mA and $I_{B2} = 0.05$ mA.

It is apparent that the operating part of the load line must be near the bottom end of the curve, since T_2 requires a base current that is very much less than $I(= 0.4$ mA). In Fig. 4.31a the operating part is indicated by the line XY, and as the voltage V_{C1} changes by only a small amount, R_{in} (the slope of XY) remains relatively constant. Generally, R_{in} will be much less than R_{C1} and efficient current transfer can be effected. At low values of I_{B2} the transfer resistance decreases, as is indicated by the slope of the V_{C2}/I_{B1} curve of Fig. 4.31b, thus introducing a degree of non-linearity. Where T_2 is required to provide a voltage output, this has the effect of limiting the effective output voltage swing.

The linearity of the stage can be improved by biasing the emitter and thus moving the load line to the right. Methods of obtaining such bias are given in Chapter 1. Alternatively, as shown in Fig. 4.32a, a forward-biased junction diode will provide a constant voltage of approximately 0.2 V which is usually adequate.

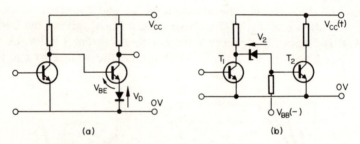

Fig. 4.32. A forward-biased diode (a) provides emitter bias which has the effect of shifting the Fig. 4.31 load line to the right, thus improving linearity. A similar effect is achieved in (b) by the use of a zener diode.

Figure 4.32b shows how a zener diode may be used to provide a low-resistance voltage drop which does not unduly increase R_{in} for T_2. It may be replaced by a resistor through which a constant current is drawn.

An approximate small signal equivalent network is shown in Fig. 4.33.

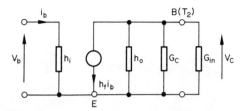

Fig. 4.33. Approximate small signal equivalent network for the evaluation of performance of a direct coupled stage. Normally, G_{in} is much greater than h_o and G_c which may therefore be neglected.

Complementary devices

With a *pnp* device, it is possible to use an *npn* as the next stage as in Fig. 4.34. This arrangement is particularly advantageous when the required output is about zero volts.

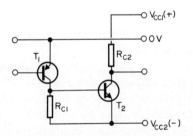

Fig. 4.34. Use of a complementary pair of transistors to provide an output voltage varying about zero.

T_2 will be cut off when T_1 is cut off. Care must be taken to ensure that the leakage current of T_1 does not produce so large a voltage across R_{C1} that T_2 is permanently bottomed. If V_{BE2} is of the order of 200 mV, R_{in} for T_2 will be low, and if R_{C1} is much greater than R_{in} the leakage current I_{CEO} will flow into the base of T_2. If T_1 has a large leakage current either R_{C1} must be kept small or a base resistor for T_2 should be used. Both these methods reduce current gain.

Design Example 4.6

Required, a d.c. amplifier with voltage gain greater than 5000 and providing signal inversion. The output resistance should be less than 1 kΩ and the output voltage swing ±5 V.

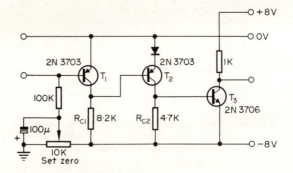

Fig. 4.35. Circuit diagram of Design Example 4.6. The 10-kΩ potentiometer enables the output voltage to be set to zero.

The circuit will take the form shown in Fig. 4.35.

Supply voltages. The output voltage is required to be ±5 V. Let the supply voltage be ±8 V. The output transistor must have $V_{CE(max)}$ greater than 16 V.

Transistors. 2N 3703's are suitable for direct coupling. The minimum current gain of two of these is $A_1 = 30 \times 30 = 900$.

As overall signal inversion is required, an odd number of common emitter stages is necessary, and an *npn* transistor with adequate collector voltage and power rating is required for the output stage. A 2N 3706 would be suitable as it has a 30-V collector rating, 350 mW dissipation at 25°C and current gain greater than 20.

Output load resistor. A 1 kΩ resistor will ensure that the output resistance is less than 1 kΩ. With this load resistor current gain $A_2 \doteqdot 20$.

$$P_{c(max)} = 8 \text{ V} \times 8 \text{ mA} = 64 \text{ mW}.$$

Gain. Voltage gain $A_v = A_i(R_L/R_{in})$,

where $A_i = A_1 A_2 \doteqdot 900 \times 20 = 18 \times 10^3$.

If $R_L = 1 \text{ k}\Omega$ and $R_{in} = 2 \text{ k}\Omega$ (approximately equal to h_i), then $A_v = -9000$.

This figure ensures that a higher value of R_{in}, or reasonable loading of R_{C3} by an external load, will not cause the gain to fall below the specified value.

R_{C1} *and* R_{C2}. A suitable current for the direct coupled pair is 1 mA.

$$R_{C1} \doteqdot \frac{V_{CC}}{I_C} = 8.2 \text{ k}\Omega \quad \text{(a preferred value)}.$$

As R_{C2} is a compromise between the current transfer and the voltage produced by leakage current, 4.7 kΩ is a suitable value.

Biasing of T_2. To improve linearity, the emitter of T_2 is held at −0.6 V by a forward-biased silicon diode, as shown in Fig. 4.32a.

Zero control. A base current control is required for T_1, to enable the output voltage to be set to zero.

4.13. Drift in transistor d.c. amplifiers

The input-referred current drift can be expressed as:

$$I_{DI} = \Delta I_{CBO} + \frac{\Delta V_{BE}}{R_s} \qquad + I_B \cdot \frac{\Delta h_F}{h_F} \qquad (4.37)$$

= leakage	+ change in V_{BE}	+ change in
current,	required to	h_F.
	maintain constant	
	I_C,	

(a)   (b)   (c)

(a) The leakage current is a function of temperature. Use of silicon transistors greatly reduces I_{CBO}.

(b) As the temperature is increased, a decrease in V_{BE} is required to maintain a constant collector current. A change of 2–4 mV per °C is commonly quoted.

It is apparent that increasing the source resistance decreases the effect of ΔV_{BE} which becomes zero for a current-fed transistor. If the amplifier has to operate from a low-resistance source, ΔV_{BE} can be balanced out by using a compensating system. Often the most satisfactory method is a *longtail pair* or symmetrical emitter-coupled system.

Both transistors should be kept at the same temperature, otherwise, even for identical units, drift would result.

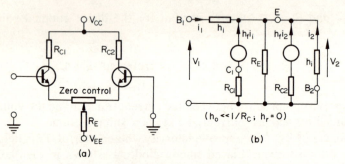

Fig. 4.36. Transistor longtail pair with equivalent network.

Longtail pair analysis

The equivalent network is simplified by considering $R_{C1} = R_{C2} = R_C$. Assume identical transistors, $h_{re} = 0$ and $h_{oe} \ll 1/R_C$ as in Fig. 4.36b.

Writing the equations for nodes B_1 and E,

$$\frac{1}{h_i} v_1 - \frac{1}{h_i} v_2 = i_1, \tag{4.38}$$

$$-\frac{1}{h_i} v_1 + v_2 \left[\frac{2}{h_i} + \frac{1}{R_E} \right] = h_f i_1 + h_f i_2. \tag{4.39}$$

From eqn. (4.38),

$$i_1 = \frac{v_1 - v_2}{h_i} \quad \text{and} \quad i_2 = \frac{-v_2}{h_i},$$

and substituting in eqn. (4.39),

$$-\frac{1}{h_i} (1 + h_f) v_1 + \frac{1}{h_i} \left(2 + 2h_f + \frac{h_i}{R_E} \right) v_2 = 0.$$

Thus,

$$v_2 = \frac{1 + h_f}{2(1 + h_f) + (h_i/R_E)} \cdot v_1, \tag{4.40}$$

$$\doteqdot \tfrac{1}{2} v_1 \quad \text{as } h_i/R_E \ll 2(1 + h_f).$$

The input current,

$$i_1 = \frac{v_1}{2h_i},$$

and

$$r_{in} = \frac{v_1}{i_1} = 2h_i.$$

The output voltage at C_1,

$$v_{C1} = -h_f R_C i_1 = -\frac{h_f}{2h_i} R_C v_1,$$

$$= \frac{-g_m R_C}{2} v_1. \tag{4.41}$$

The T_2 base current,

$$i_2 = -\frac{v_2}{h_i} = -\frac{v_1}{2h_i}.$$

The output voltage at C_2,

$$v_{C2} = h_f R_C i_2 = \frac{h_f}{2h_i} R_C v_1,$$

$$= \frac{g_m R_C}{2} v_1. \tag{4.42}$$

Note that half the input voltage v_1 is across the input transistor and half across the emitter resistor R_E.

DESIGN EXAMPLE 4.7

Required, a longtail pair to provide a voltage gain greater than 40, using supply voltages of ±5 V.

Transistors. Let the transistors be 2N 3706 types with a g_m greater than 30 mA/V (i.e. 30 mS).

Collector resistor R_C

$$A_v = 40 = \frac{g_m R_C}{2}, \quad \text{from eqn. (4.42).}$$

Thus,

$$R_C = \frac{2A_v}{g_m} = \frac{2 \times 40}{30 \times 10^{-3}} = 2.7 \text{ k}\Omega.$$

Let the standing voltage of the collector for zero input be 2.5 V. This allows a collector swing of ±2.5 V about the quiescent value:

$$I_C = \frac{V_C - V_{CC}}{R_C} = \frac{2.5 \text{ V}}{2.7 \text{ k}\Omega} \doteq 1 \text{ mA.}$$

Emitter resistor R_E

$$R_E \doteq \frac{V_{EE}}{2I_C} = \frac{5 \text{ V}}{2 \text{ mA}}.$$

Let it be 2.7 kΩ which is a preferred value.

Dissipation. 2.5 V × 1 mA = 2.5 mW which is well within the transistor rating.

The final circuit details are as shown in Fig. 4.37.

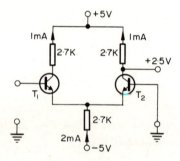

Fig. 4.37. Circuit diagram of Design Example 4.7.

Difference amplifier

In Fig. 4.38a, instead of taking the base of T_2 to earth, a second signal v_2 is introduced to the base. Then, due to v_1, from eqn. (4.42),

$$v'_{C2} = \frac{g_{m1}R_C}{2} v_1 = A_1 v_1, \tag{4.43}$$

where $A_1 = g_{m1}R_C/2$ is the gain from base 1 to collector 2. Also, due to v_2, from eqn. (4.41)

$$v''_{C2} = -\frac{g_{m2}R_C}{2} v_2 = A_2 v_2, \tag{4.44}$$

where $A_2 = -g_{m2}R_C/2$ is the gain of T_2.

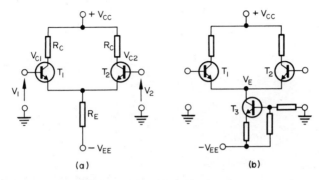

Fig. 4.38. (a) The longtail pair as a difference amplifier. (b) Use of a constant current tail to reduce common mode gain.

These combine to yield,

$$v_{C2} = \frac{g_{m1}R_C}{2} v_1 - \frac{g_{m2}R_C}{2} v_2 \qquad (4.45)$$

If $g_{m1} = g_{m2}$,

$$v_{C2} = \frac{g_m R_C}{2} (v_1 - v_2),$$

$$= A_d v_d,$$

where

$$A_d = \tfrac{1}{2}(A_1 - A_2) \qquad (4.46)$$

is the overall gain of the difference amplifier, and the differential input voltage is

$$v_d = v_1 - v_2. \qquad (4.47)$$

The output is thus a function of the difference between the two input signals and if, in this case, $v_1 = v_2$, the output is zero.

Common mode effects

In practice g_{m1} does not exactly equal g_{m2}, and the output is not only a function of v_d, but also of v_{cm}, the average level of the two signals. v_{cm} is called the *common mode* signal,

$$v_{cm} = \tfrac{1}{2}(v_1 + v_2). \qquad (4.48)$$

Equation (4.45) may be written as

$$v_{C2} = A_1 v_1 + A_2 v_2. \qquad (4.49)$$

Also, from eqns. (4.47) and (4.48),

$$v_1 = v_{cm} + \tfrac{1}{2} v_d,$$

and

$$v_2 = v_{cm} - \tfrac{1}{2} v_d.$$

Substituting in eqn. (4.49)

$$v_{C2} = v_{cm}(A_1 + A_2) + v_d \tfrac{1}{2}(A_1 - A_2),$$
$$= A_c v_{cm} + A_d v_d. \qquad (4.50)$$

where A_c, the common mode gain, is

$$A_c = A_1 + A_2.$$

It is obviously desirable to obtain the maximum signal gain and to minimize the common mode gain, and a figure of merit by which amplifiers may be compared is the *common mode rejection ratio* (CMRR) $\rho = A_d/A_c$.

Effect of R_E

If equal signals of opposite sense are applied to the two inputs of the longtail pair, they cause equal but opposite changes to the emitter current flowing through R_E. The emitter voltage V_E remains unchanged and no loss in signal gain results. Common mode signals, since they act in the same sense at both inputs, however, cause emitter current changes which are additive. Thus, if with *npn* transistors the common mode input signals go positive, I_E increases and V_E goes positive. This causes V_{BE} to be reduced, which acts to cancel out the effect of the incoming signal. The common mode gain is thus reduced by the action of current negative feedback. To an approximation the common mode gain is given by $A_c \doteq -R_C/2R_E$ so a high value of R_E is indicated. There is, however, a limit to how much it can be increased. For as it is made bigger, so must be V_{EE} in order to maintain the same quiescent current. Alternatively if the quiescent current is allowed to decrease, h_{fe} is decreased and, with it, the common mode rejection ratio. A solution is to replace R_E by a so-called *constant current* tail, as shown in Fig. 4.38b. For T_3 the base emitter voltage V_{BE} is independent of the signal voltages so, for a given collector voltage, the current is constant. Thus, with equal but opposite sense input signals, the common emitter voltage remains essentially constant, as does I_E. However, with common mode signals going positive, so does the common emitter voltages, and I_E increases, giving current negative feedback as before. The high value of dynamic resistance of T_3 gives a very much reduced common mode gain. Additionally, since transistors can be made more cheaply than resistors, this is the circuit form used in the differential input circuits of all integrated circuit d.c. amplifiers.

The use of a differential input tends to reduce drift in d.c. amplifiers since drift sources are largely common mode. An additional feature of this input stage is that two input terminals are available. Referring to eqns. (4.43) and (4.44), an input at one terminal gives rise to an output signal which is nominally in phase with it, the *non-inverting input*. The other input, the *inverting input*, gives rise to an output signal of the opposite sense.

Differential amplifiers having high common mode rejection find wide use in the medical electronics field, for instance in the measurement of electromyographic (EMG) signals, which occur during muscular activity. Two electrodes are used to pick up opposite sense signals of μV strength, in the presence of high noise level signals. When applied to the amplifier, the EMG signals are amplified while the noise signals, which are common mode, i.e. the same at both input terminals, are largely rejected.

4.14. Integrated circuit amplifiers

The majority of integrated circuit and modular amplifiers currently available are of the zero frequency type. The impetus for the development of these has come from the analogue computing and control engineering fields of application, and they are commonly referred to as *operational amplifiers*. Various types of "op-amps" may be obtained, but it is unreasonable to expect that any device will exactly suit the requirements for a given application. Accordingly, it is the task of a systems engineer to select an amplifier, and then to modify its performance by feedback and other means, using external circuitry. Selection is made on the basis of a number of characteristics, some of which are now explained.

4.15. Operational amplifier characteristics

Ideally, an operational amplifier will have infinite gain, infinite input impedance and zero output impedance. The degree to which these parameter values are achieved determines the accuracy with which a unit can be used for analogue computing purposes, for instance, as is explained in § 6.5. These, and other characteristics listed below are the worst case values usually quoted in manufacturers' data sheets.

Open-loop gain

This is the voltage gain of the amplifier at d.c. or low frequency, with no feedback components connected. It is commonly stated in decibels, i.e. $20 \log_{10}(V_{out}/V_{in})$, and gains of 90 to 100 dB are typical. The use of the decibel to express voltage gain, however, must be qualified. The dB basically expresses gain as the ratio of two powers, i.e.

$$\text{Gain} = 10 \log_{10} (P_o/P_{in}) \text{ dB},$$

where P_o is the output power of a device and P_{in} is the input power. If P_{in} watts is expressed as V_{in}^2/R_{in} and P_o watts is expressed as V_o^2/R_o, then

$$\text{Gain} = 10 \log_{10} \left[\frac{V_o^2}{R_o} \cdot \frac{R_{in}}{V_{in}^2} \right] \text{ dB},$$

and if $R_o = R_{in}$, this can be written as

$$\text{Gain} = 20 \log_{10} \left(\frac{V_o}{V_{in}} \right) \text{ dB}.$$

Thus, to express voltage gain in decibels implies that the two impedances are equal.

Input impedance

Manufacturers quote the dynamic resistance and shunting capacitance, measured at one input terminal with the other terminal earthed. This is the *differential input impedance.* For bipolar devices typical values are in the range 300 kΩ to 10 MΩ shunted by 3 pF, while for units employing FET input stages, input resistances better than $10^{11}\Omega$ are available. The *common mode input resistance* is the dynamic resistance measured between the two inputs connected together, and earth, and is of much higher value.

Input offset voltage

If one input is earthed, this is the voltage which must be applied to the other input terminal in order to obtain zero output. The input offset voltage is quoted in mV at 25°C and is a function of temperature and the power-supply voltage. Offset voltage temperature drift is stated in $\mu V/°C$ and is typically in the range ±5 to ±50, although in chopper stabilized units drifts as low as 0.1 $\mu V/°C$ are quoted. The *power supply rejection ratio* is the ratio of the change in input offset voltage to the change in power-supply voltage producing it.

Most differential input operational amplifiers have provision for compensating initial offset voltage to zero using an external trim potentiometer.

Input bias current

This is the average value of currents into the two input terminals when the output voltage and the mean input voltage are both zero. Alternatively, the initial *input offset current* is the difference between the two input currents for the same conditions. The parameter is of significance in the case where input signals are applied via high value resistors, and is again a function of temperature. Bias currents are usually quoted in nA at 25°C, accompanied by an indication of the change with temperature stated in nA/°C.

Input voltage range

The range of voltages at the input terminals, for which the amplifier performs according to specification. This is stated in two forms. The *differential input voltage* specifies the maximum voltage which can be applied between the two input terminals without causing permanent damage to the amplifier. The *common mode input voltage* range specifies the range of common mode voltages which can be accommodated whilst still permitting small differential input signals to be satisfactorily amplified.

Note that when operating with a high degree of feedback (see

Chapter 6), if an excessive voltage is applied to the inverting input, certain types of amplifiers are subject to *gain reversal*. This results in a "latch-up" condition, in which the output remains at one of its limits until power is switched off.

Output impedance

This is the dynamic resistance measured at the output terminal and for bipolar circuits is typically $2 \, k\Omega$.

Rated output

Although high voltage–high current units are available, the majority of operational amplifiers provide an output of $\pm 10 \, V$ with current in the range 5 to 20 mA. This defines the maximum peak to peak output voltage swing which can be obtained without signal "clipping", when the amplifier is properly zeroed. Where greater output current is necessary, *power boosters* are available, which are capable of providing up to 0.5 A at $\pm 10 \, V$. These are employed in conjunction with low power op-amps, for use as power output stages.

An additional specification, the *output short-circuit current*, sets a limit to the current which can be safely drawn from the amplifier when the output is short circuited either to earth or to one of the supply lines.

Frequency response

Since an amplifier can amplify down to zero frequency, a quoted bandwidth is essentially a statement of an upper frequency limit. The various forms in which this is given are indicated in Fig. 4.39 and are as follows. The *small signal 3 dB bandwidth* states the frequency at which the open-loop voltage gain is reduced to 0.707 of its d.c. value, and the *unity gain bandwidth*, the frequency at which the open-loop gain is reduced to unity. The *full power bandwidth* is measured under unity closed-loop gain conditions, and indicates the maximum frequency for which the maximum output voltage swing of a sinusoidal signal may be obtained without distortion.

Another indication of the maximum frequency an amplifier can handle is its *slew rate*. This states the maximum rate of change of output voltage under maximum output signal conditions. Note that

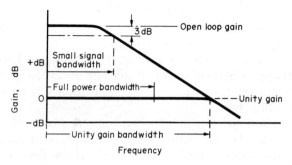

Fig. 4.39. An illustration of the terminology used to describe the frequency response of operational amplifiers. The full power bandwidth indicates the highest frequency that maximum output can be obtained without distortion.

this is not necessarily the same in both directions. The slew rate obviously influences the way in which an amplifier can respond to a step input signal of large amplitude.

Transient response

The curve of Fig. 4.40 is typical of the way a closed-loop system might respond to a step input, and serves to illustrate the terminology used to describe the transient performance of an amplifier. *Rise time* is the time required for an output voltage step to change from 10 per cent to 90 per cent of its final value, and to some extent is dependent upon the magnitude of the input step signal. For large signal swings at or near unity gain the output voltage will rise at the amplifier's specified slew rate, overshoot, and then settle rapidly.

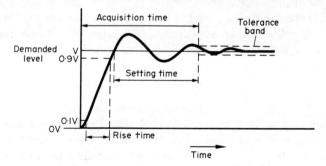

Fig. 4.40. Transient response of an operational amplifier.

For small signal swings, however, when only a small part of the output dynamic range is utilized, the step response is mostly limited by the small signal frequency response.

Acquisition time.   This is a measure of the delay between the time of the application of the input step and the instant at which the output finally settles within a specified tolerance band about the demanded final level.

Settling time.   This is the time between the instant at which the output first reaches the demanded level and that at which it finally settles within the specified tolerance band about that level.

Common mode rejection ratio

This has previously been defined as the ratio of common mode gain to the differential signal gain, and is a measure of the ability of an amplifier to reject common mode signals. It can otherwise be defined as the ratio of a specified change of common mode input voltage to the associated change of differential input voltage required to maintain the output voltage unchanged, and is expressed in decibels.

Note that for an inverting amplifier, the input terminal is at *virtual earth*, and sees only a very small error signal. Thus there is no common mode signal and the common mode rejection specification is irrelevant.

4.16.   Types and applications

The three amplifier characteristics of principle importance are drift, input impedance and speed of response, and, in general, any one of these can be improved at the expense of the others. To assist in selection for a given application, integrated circuit and modular operational amplifiers can therefore be classified into four groups, as follows.

General purpose

Amplifiers in this group are used where an overall moderate performance is acceptable. The first widely used general purpose device was the type 709. This normally required the use of *anti-latch-up* diodes at the input, a *C–R* compensation network for stable operation, and a series output resistor to limit current. It has now been replaced by the 741 which needs no external circuitry and may be used for integrator, summing amplifier and general feedback applications.

When fed from a source resistance of less than 10 kΩ the 741 has a maximum input offset voltage drift of 15 μV/°C, an input offset current drift of 0.5 nA/°C, a typical input impedance of 6 MΩ and a bandwidth of about 1 MHz. For a ±15-V power supply, when feeding a load greater than 2 kΩ the device provides a large signal voltage gain greater than 50,000. The slew rate, however, is a modest 0.5 V/μs which limits the usefulness of this low cost device in many applications.

By comparison, general purpose bipolar amplifiers of modular construction have slew rates up to 2 V/μs, with bandwidths up to 2 MHz unity gain and 20 kHz full power. Their drift rates are such that best performance is obtained when they are used in situations where external circuit impedances are less than 50 kΩ.

High input impedance

Where a signal is obtained from a high impedance source (above 100 kΩ) the input bias current should be very low and, to avoid loading the source, the input impedance of the amplifier should be very high. Amplifiers which meet this requirement normally have either FET or varactor bridge input stages and are used for high-accuracy integrators, sample and hold circuits and current-to-voltage amplifiers. An input impedance of 10^{12} Ω is typical and FET amplifiers operate with bias currents in the range 0.1 to 1.0 pA. Varactor amplifiers provide an even lower bias current characteristic (0.01 pA), and additionally have a very low value of input noise. This latter feature makes them suitable for the amplification of signals from such sources as photo-multiplier tubes and radiation detectors, examples in which the ability to amplify very small signals is of paramount importance. The bandwidth of the varactor amplifier, however, is limited to about 2 KHz unity gain and 7 Hz full power.

Low drift

For the amplification of low-level signals, and for highly accurate

integration, it is important that the effects of input offset voltage and
current drift should be minimal. In selecting an amplifier for such
applications, it is important to take account of the effect of source
resistance.

The ratio of input offset voltage drift to current drift yields a value
of resistance. For a source resistance less than this value, the major
contribution to overall drift is due to the input offset voltage of the
amplifier itself. For higher values of source resistance, however, the
voltage dropped by the input offset current flowing through the
source is significant, and current drift is then the dominant
parameter.

Low drift amplifiers are manufactured in bipolar, FET and
chopper stabilized forms. The best of these, and the most costly, is
the chopper stabilized type. This provides a very low value of input
offset voltage drift, combined with an extremely high open loop
gain. Chopper amplifiers are normally single ended input devices,
however, and are only applicable as inverting amplifiers.

High speed

This fourth group of amplifiers includes those in which the design
emphasis has been on achieving an extremely large bandwidth and
fast response, and generally refers to devices having unity gain
bandwidths in excess of 10 MHz.

One of the largest areas of use for high-speed operational
amplifiers is in data conversion, and in a digital to analogue conver-
ter, speed of conversion is limited by how fast the amplifier can
respond to its input signal. The ability of an amplifier to respond
rapidly is determined not only by its bandwidth but also by its slew
rate and settling time, and selection of a device for such an
application should be made on the basis of all three parameters and
not just one alone.

In this respect it should be remembered that slew rate is, to some
extent, dependent upon the values of capacitors used for stabilization purposes. Since these values are often different for the inverting
and non-inverting applications, two values of slew rate should be
considered. Additionally, slew rate is a function of the closed-loop
gain in use, and is not necessarily the same for the leading and
trailing edges of a signal pulse.

The question of settling time is also not straightforward, since the definition of this parameter is not standard among manufacturers. Some define a settling time in such a way that it represents only a fraction of the period represented in other specifications. Fast settling amplifiers are used in high accuracy conversion systems providing, for instance, 0.1 per cent (ten-bit) accuracy or 0.02 per cent (twelve-bit) accuracy. For an amplifier of ± 10 V output this implies 10 mV and 2 mV of error, respectively, and any noise voltages in the system must be less than this. In conclusion it is fundamental that, in any closed-loop system, the greater the gain the smaller is the error. Thus, for example, to ensure that the output of a unity gain non-inverting amplifier settles within ± 0.01 per cent the amplifier gain should be greater than 10,000.

CHAPTER 5

Tuned Amplifiers

INTRODUCTION

The tuned amplifier is used when it is required to amplify h.f. signals at one frequency, or band of frequencies, and reject others. A range of frequencies between 150 kHz and 50 MHz is included in this category of amplifiers. Selectivity is obtained by the use of a parallel L–C circuit which resonates at the desired frequency and which, at this frequency has the high impedance necessary for the load of a voltage amplifier. The important characteristics of a tuned amplifier are the gain at resonance, the variation of gain with frequency in the immediate vicinity of resonance, and, if the frequency is to be varied, the way in which the gain changes when this is done.

5.1. The parallel-tuned circuit

In Fig. 5.1a a parallel-tuned circuit is drawn, made up of an inductance L having resistance r, and shunted by a capacitance C.

Resonance will occur when the capacitive and inductive reactances are equal, i.e. when $\omega_0 L = 1/\omega_0 C$, where $\omega_0 = 2\pi f_0$. From

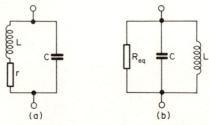

(a)   (b)

Fig. 5.1. Parallel-tuned circuit with its resistive component (a) in series and (b) in parallel.

this, the resonant frequency,

$$f_0 = \frac{1}{2\pi\sqrt{(LC)}}, \qquad (5.1)$$

and the magnification

$$Q = \frac{\omega_0 L}{r} = \frac{1}{\omega_0 C r} = \frac{1}{r}\sqrt{\left(\frac{L}{C}\right)}. \qquad (5.2)$$

The impedance of such a circuit may be written

$$Z(j\omega) = \frac{(r + j\omega L)/j\omega C}{r + j(\omega L - 1/\omega C)}, \qquad (5.3)$$

and

$$Z(j\omega) = \frac{L/C}{r + j(\omega L - 1/\omega C)},$$

if r is much less than ωL. At resonance, when $\omega_0 L = 1/\omega_0 C$,

$$Z = \frac{L}{Cr} = R_D. \qquad (5.4)$$

The impedance at resonance is purely resistive and is known as R_D, the *dynamic resistance*. The network of Fig. 5.1a may be redrawn as in Fig. 5.1b in which C and L are shunted by R_{eq}. The two networks are equivalent at resonance provided that $R_{eq} = L/Cr$. From this, $r = L/CR_{eq}$ and if this is substituted in expression (5.2),

$$\text{magnification} \quad Q = R_{eq}/\omega_0 L = \omega_0 C R_{eq} = R_{eq}\sqrt{\left(\frac{C}{L}\right)}. \qquad (5.5)$$

5.2. Single-tuned circuit amplifier

A simplified circuit diagram is given in Fig. 5.2a, the coupling capacitor C_C serving to prevent the T_1 drain d.c. voltage appearing at the gate of T_2. The equivalent network of this circuit is as shown in Fig. 5.2b. C represents the total circuit capacitance and includes the input capacitance of the second transistor, and stray capacitance associated with wiring, etc. which together with R_G (the input resistance of T_2) are effectively in parallel with the tuned circuit.

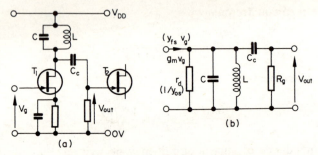

Fig. 5.2. Basic arrangement of a single-tuned circuit amplifier with its equivalent network. C_c is a blocking capacitor which keeps the T_1 drain voltage off the gate of T_2.

The value of C_C is chosen such that it has negligible reactance at the frequencies being amplified, and as it does not shunt the tuned circuit it may be neglected.

Design considerations

Examination of the equivalent network shows that the drain resistance of the FET is effectively in parallel with the tuned circuit load. A transistor with a high r_d is therefore required to prevent damping of the tuned circuit. As the impedance of the load will be high, large gains are possible and difficulties due to feedback may be encountered. Hence it is important that drain to gate capacitance be low if instability is to be avoided.

The equivalent circuit is redrawn in Fig. 5.3, in which R represents the total circuit resistance including r_d of T_1, the input resistance of T_2 and the equivalent shunt loss resistance of L and C.

The impedance of the tuned circuit may be written,

$$Z(j\omega) = \frac{LR/C}{L/C + j(\omega LR - R/\omega C)}. \qquad (5.6)$$

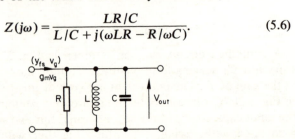

Fig. 5.3. Simplified equivalent network of the drain circuit of Fig. 5.2.

Let d, the dissipation factor, equal $1/Q$, thus,

$$d = \frac{1}{\omega_0 CR} = \frac{\omega_0 L}{R} = \frac{1}{R}\sqrt{\left(\frac{L}{C}\right)}.$$

From eqn. (5.6),

$$Z(j\omega) = R\left\{\frac{(1/R^2)(L/C)}{(1/R^2)(L/C) + j[(\omega L/R) - (1/\omega CR)]}\right\},$$

$$= R\left\{\frac{d^2}{d^2 + jd[(\omega LR/\omega_0 LR) - (\omega_0 CR/\omega CR)]}\right\},$$

therefore

$$Z = R\left\{\frac{d}{d + j[(f/f_0) - (f_0/f)]}\right\}, \tag{5.7}$$

where f_0 is the resonant frequency and f is any other frequency.

The gain of the circuit from the gate of T_1 to the gate of T_2 is $g_m Z$, and is maximum at resonance when $Z = R$. At frequencies off resonance the impedance is less than R and the gain is reduced as shown in Fig. 5.4.

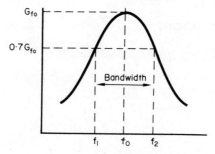

Fig. 5.4. Amplitude response of a single tuned-circuit amplifier. The lower and upper 3 dB points are f_1 and f_2.

As previously, let the bandwidth be defined as the range of frequencies between the two points at which the gain is 3 dB down, i.e. $1/\sqrt{2}$ times the gain at resonance. Referring to eqn. (5.7), this occurs when $(f/f_0) - (f_0/f) = \pm d$.

For circuits having a low dissipation factor f will be close to f_0, and the gain is 3 dB down when,

$$\frac{f^2 - f_0^2}{f_0^2} \doteq d,$$

and

$$\frac{f - f_0}{f_0} \cdot \frac{f + f_0}{f_0} = d.$$

To a close approximation

$$\frac{f + f_0}{f_0} = 2.$$

Hence, writing $\Delta f_0 = f - f_0$,

$$\frac{2\Delta f_0}{f_0} = d \quad \text{and} \quad \Delta f_0 = \frac{f_0 d}{2}.$$

The gain is therefore 3 dB down at the two frequencies

$$f_1 = f_0 - \frac{f_0 d}{2}, \quad f_2 = f_0 + \frac{f_0 d}{2}. \tag{5.8}$$

From these expressions,

$$d = \frac{f_2 - f_1}{f_0} = \frac{\text{bandwidth}}{f_0},$$

and since

$$d = \frac{1}{2\pi f_0 RC},$$

$$\text{bandwidth} = \frac{1}{2\pi RC} = \frac{f_0}{Q}. \tag{5.9}$$

For any circuit configuration the gain–bandwidth product is a constant, it being possible to increase bandwidth at the expense of gain and vice versa. In the case of a single tuned circuit stage where gain $= g_m Z$ ($= g_m R$ at resonance),

$$\text{gain–bandwidth product} = g_m / 2\pi C \text{ Hz}. \tag{5.10}$$

Design steps
1. Select a suitable operating point, and determine the value of R_S the source resistor.
2. Decouple R_S.
3. For a given gain evaluate the necessary effective R.
4. Substitute this value of R in eqn. (5.9) and thus determine the capacitance required in order to obtain the correct bandwidth. By subtracting the sum of the stray capacitances from this, the value of the tuned circuit capacitor is obtained.
5. From eqn. (5.1) evaluate L which, with C, will resonate at the desired frequency.
6. Calculate the dynamic resistance of this tuned circuit and determine what shunting resistance is required to provide the correct value of R_{eff} as found in Step 1. To do this a typical coil resistance is assumed.

DESIGN EXAMPLE 5.1

Required, an amplifier with a gain of 100 at a frequency of 200 kHz. The bandwidth is to be 10 kHz, and a supply voltage of 8 V is available.

Operating point. Neglecting the ohmic resistance of the coil, which is small, the drain will have a standing voltage of the full $V_{DD} = 8$ V. For the selected FET let the operating point be $V_{DS} = 8$ V, $I_D = 2$ mA and $V_{GS} = -1$ V, and, under these conditions let $g_m = 2$ mS and $r_d = 200$ kΩ.

$$R_S = 1 \text{ V}/2 \text{ mA} = 500 \text{ Ω}.$$

At 200 kHz a 0.1-μF capacitor has an impedance of less than 1 Ω, so this is suitable for decoupling R_S.

Tuned circuit load.

$$R_{eff} = \text{gain}/g_m = 100/(2 \times 10^{-3}) = 50 \text{ kΩ}.$$

From eqn. (5.9), bandwidth $= 1/2\pi RC = 10$ kHz. Thus,

$$C = \frac{10^{-6}}{2\pi \times 50 \times 10} = 318 \text{ pF}.$$

Let the tuned circuit capacitor $C = 290$ pF which allows 28 pF for stray capacitance.

From eqn. (5.1),

$$L = \frac{1}{4\pi^2 f_0^2 C} = \frac{1}{40 \times 4 \times 10^{10} \times 290 \times 10^{-12}}.$$

Thus,

$$L \doteqdot 2\,\text{mH}.$$

Assuming a typical coil resistance of 30 Ω,

$$R_D = \frac{L}{Cr} = \frac{2 \times 10^{-3}}{290 \times 10^{-12} \times 30} = 230\,\text{k}\Omega,$$

and, referring to the equivalent network of Fig. 5.2b, this is in parallel with r_d and R_G. A reasonable value for R_G would be 1 MΩ, so for $r_d = 200\,\text{k}\Omega$ the effective resistance is approximately 97 kΩ.

However, to satisfy the gain and bandwidth conditions it was earlier determined that R_{eff} should be 50 kΩ. It is therefore necessary to shunt the tuned circuit with a resistor such that,

$$\frac{1}{R} = \frac{1}{50\,\text{k}\Omega} - \frac{1}{97\,\text{k}\Omega}, \quad \text{and } R \doteqdot 112\,\text{k}\Omega.$$

Let it be 110 kΩ.

Coupling capacitor. This is merely required as a blocking capacitor to keep the T_1 drain voltage off the gate of T_2. It should have a reactance which is negligible at the frequency to be amplified, and a 0.001 μF capacitor will be suitable. This completes the circuit, which is drawn in Fig. 5.5.

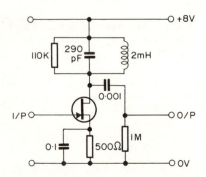

Fig. 5.5. Completed circuit of Design Example 5.1.

5.3.   Tunable r.f. amplifier with constant selectivity[37]

Where it is required that the frequency of an r.f. amplifier be tunable, it is usually desirable that, over the tuning range, the overall selectivity should remain constant.

The resonant frequency of a tuned circuit may be changed by varying either the inductance or the capacitance of the circuit. With permeability tuning, the inductance is varied by adjusting the position of an iron dust core within the coil. This produces less variation in selectivity and gain over the tuning range than does capacitor tuning, but is both more costly and more difficult to achieve. Capacitor tuning is therefore the more commonly used method, and with it special arrangements must be made in order to obtain constant selectivity.

Design considerations
In eqn. (5.9) the bandwidth of a parallel tuned circuit was given as

$$B = f_0/Q.$$

For constant selectivity, bandwidth should be constant and hence Q should be proportional to f_0. Since $Q = \omega L/r$ it would appear that, with a fixed inductance having a given ohmic resistance, Q is proportional to f_0. In fact, as frequency increases, the resistance of the coil also increases and, particularly with iron-cored coils, Q decreases as shown in Fig. 5.6. It is therefore necessary to make use

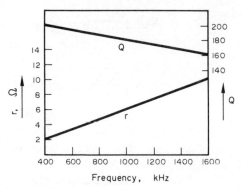

Fig. 5.6. The resistance and Q of an inductance as a function of frequency.

of a network having a resistive component decreasing with frequency, which when added to the coil resistance will make the total resistance constant with frequency.

In Fig. 5.6, in which r increases from 2 Ω to 10 Ω as f increases from 400 kHz to 1600 kHz, there is needed a network whose resistance falls from 8 Ω to zero over the same frequency range causing the total resistance R to remain constant at 10 Ω. Such a technique has the disadvantage that it derates the circuit performance at the low frequencies, but the required network which is given in Fig. 5.7 is a simple one. Let the reactance of the L–C arm be X.

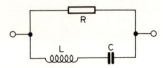

Fig. 5.7. Network used to obtain constant selectivity in a tuned circuit.

Then,

$$Z - \frac{jRX}{R + jX},$$

$$= \frac{RX^2 + jR^2X}{R^2 + X^2}.$$

Resistive component

$$R_1 = \frac{RX^2}{R^2 + X^2}. \tag{5.11}$$

Reactive component

$$X_1 = \frac{R^2X}{R^2 + X^2}. \tag{5.12}$$

The values of L and C may be chosen such that at the highest frequency in use the reactance is zero. Then, from eqn. (5.11), the resistive component R_1 is also zero. As the frequency decreases X increases, and so therefore does R_1.

The network thus has the required characteristic and is used in series with the inductance of the amplifier tuned circuit. Its reactive component is usually small and may be neglected in comparison to the reactance of the coil itself.

The impedance of the L–C arm is

$$jX = j\omega L - \frac{1}{j\omega C} = j\omega L \left(1 - \frac{1}{\omega^2 LC}\right). \qquad (5.13)$$

But from eqn. (5.11), $R_1 = 0$ when $X = 0$, that is at the resonant frequency of the network.

Then,

$$\omega_0 L = \frac{1}{\omega_0 C} \quad \text{and} \quad \omega_0^2 LC = 1,$$

therefore

$$jX = j\omega L \left(1 - \frac{\omega_0^2 LC}{\omega^2 LC}\right) \quad \text{and} \quad X = \omega L \left(1 - \frac{f_0^2}{f^2}\right). \qquad (5.14)$$

Substituting in eqn. (5.11),

$$R_1 = \frac{R\{\omega L[1 - (f_0^2/f^2)]\}^2}{R^2 - \{\omega L[1 - (f_0^2/f^2)]\}^2}, \qquad (5.15)$$

and rearranging,

$$L^2 = \frac{R_1 R^2}{\omega^2 (R - R_1)[1 - (f_0^2/f^2)]^2}. \qquad (5.16)$$

There are three unknowns, L, C and R, so values for R_1 are selected from Fig. 5.6 for three frequencies and substituted in the above equations.

Design steps

1. Set up the bias conditions as in Design Example 5.1.
2. Calculate the maximum and minimum values of capacitance associated with the tuning capacitor and circuit, and thus determine the inductance for the frequency range to be covered.
3. Experimentally obtain the value of coil resistance over the range and plot a graph as in Fig. 5.6.
4. Select three frequencies at which selectivity compensation is to be applied and, at the two lower frequencies, note the

values of R_1 required to make the total resistance, in each case, equal to the resistance of the coil at the highest frequency.
5. Substitute in eqns. (5.15) and (5.16) to determine values of L, C and R of the selectivity compensation network.

DESIGN EXAMPLE 5.2

Required, an amplifying stage covering the range 540–1600 kHz with constant selectivity.

For the selected FET let the correct bias conditions be set up as in Design Example 5.1.

The tuned circuit. A typical variable capacitor, with the stray capacitance of the circuit, will vary over a range of approximately 50–550 pF. With an inductance of 160 μH this gives a frequency coverage of 540–1800 kHz which meets the specification.

If the resistance of this coil varies with frequency as is shown in Fig. 5.6, then at 1600 kHz, $r = 10\,\Omega$ and $Q = \omega L/r = 160$. The compensation network must provide a resistive component such that the effective R is 10 Ω and $Q = 160$ over the whole frequency range. In fact this will only be achieved accurately at the three chosen frequencies but between these the error will be small. Let the chosen frequencies be 600, 1100 and 1600 kHz.

At 600 kHz, $r = 3.4\,\Omega$ and required $R_1 = 6.6\,\Omega$.
At 1100 kHz, $r = 6.7\,\Omega$ and required $R_1 = 3.3\,\Omega$.
At 1600 kHz, $r = 10\,\Omega$ and required $R_1 = 0$.

This latter condition is obtained at resonance, so $f_0 = 1600$ kHz.

f = 1100 kHz:
$$R_1 = 3.3\,\Omega \quad \text{and} \quad f_0^2/f^2 = 2.11.$$

Substituting in eqn. (5.16),

$$L^2 = \frac{3.3R^2 \times 10^{-12}}{59.5R - 196.4}\,\text{H}. \tag{5.17}$$

f = 600 kHz:
$$R_1 = 6.6\,\Omega, \quad \omega L = 3768 \times 10^3 L \quad \text{and} \quad f_0^2/f^2 = 7.1,$$

therefore,

$$\left[\omega L \left(1-\frac{f_0^2}{f^2}\right)\right]^2 = 37.2\omega^2 L^2 = \frac{1768R^2}{59.5R - 196.4}, \qquad (5.18)$$

and substituting this in eqn. (5.15) and solving, gives that $R = 7.5\ \Omega$.

Hence, from eqn. (5.17), $L = 0.86\ \mu\text{H}$. Also at the resonant frequency of 1600 kHz, $C = 1/\omega_0^2 L = 0.01135\ \mu\text{F}$.

In the completed circuit of Fig. 5.8, the tuning capacitor is taken to earth instead of actually shunting the coil and compensating network. Consideration of an equivalent circuit shows that as far as the signal is concerned the capacitor serves the same purpose when connected in this manner. There is also the advantage that hand capacity effects are avoided.

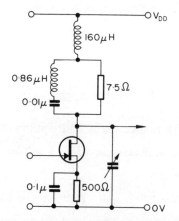

Fig. 5.8. Completed circuit of Design Example 5.2 to provide a tunable stage with constant selectivity.

5.4. Cascaded single-tuned amplifiers

The gain–bandwidth product of an amplifying stage is a figure of merit by means of which different circuit configurations may be compared. The product is a constant for a given circuit such that bandwidth may only be increased at the expense of gain. At broadcast frequencies this imposes very little limitation on design since the bandwidths in use are normally less than 10 kHz, and for stability reasons a gain of 150 is rarely exceeded in a single stage.

At much higher frequencies, however, bandwidths of the order of several megaherz are sometimes required and the gain of a single stage is consequently restricted. In order to obtain a specified gain it is therefore often necessary to use more than one stage of amplification.

When a number of identical stages are used in cascade the figure of merit for the complete amplifier is not over-all gain times over-all bandwidth, but rather (over-all gain)$^{1/n}$ × over-all bandwidth, where n is the number of stages.[36] As the number of stages is increased the over-all bandwidth decreases and, as an approximation,

$$\text{Over-all bandwidth} = \frac{\text{Single stage bandwidth}}{1.2\sqrt{n}}. \qquad (5.19)$$

The gain–bandwidth product thus becomes

$$G^{1/n}B_n = \frac{g_m}{2\pi C} \cdot \frac{1}{1.2\sqrt{n}}, \qquad (5.20)$$

so that,

$$B_n = \frac{g_m}{2\pi C} \cdot \frac{1}{1.2\sqrt{n}} \cdot \frac{1}{G^{1/n}}. \qquad (5.21)$$

Differentiating with respect to n and equating to zero gives that

$$B_n \text{ is maximum when } n = 2 \log_e G.$$

Therefore

$$G = e^{n/2}, \quad \text{and} \quad G^{1/n} = e^{1/2} = 1.65.$$

This means that for an amplifier of this type, with a given over-all gain, maximum bandwidth is obtained when the individual stage gain is 1.65 (or 4.34 dB).

Example.   Using the FET of Design Example 5.1, let the requirement be a 30 MHz amplifier with an over-all gain of 70 dB and with maximum bandwidth.

If the stray capacitance is 28 pF (as before), then the required inductance is 1.0 μH and no tuned circuit capacitor is needed.

Number of stages = 70/4.34 = 16.

$$G^{1/n}B_n = \frac{g_m}{2\pi C} \cdot \frac{1}{1.2\sqrt{n}} = 2.36 \text{ MHz}.$$

Therefore bandwidth = 2.36/1.65 = 1.43 MHz.

An increase in the number of stages beyond 16 will cause reduction in bandwidth. This example illustrates the importance of considering the gain–bandwidth product when designing multi-stage tuned amplifiers. However, field-effect transistors are available which would provide a gain-bandwidth of this magnitude with far fewer stages. For example, the wideband BFS 28 has a g_m of 12 mS. Since gain–bandwidth = $g_m/2\pi C$, this device, in a single-tuned stage, is capable of providing approximately 68 MHz and only three stages of amplification are required to obtain the same result. Thus, over-all gain = 70 dB = 3162, so for three stages, gain per stage = $^3\sqrt{3162}$ = 15. Single-stage bandwidth = 68 MHz/15 = 4.53 MHz and the overall bandwidth = $4.53/(1.2\sqrt{3})$ = 2.2 MHz.

5.5. Staggered-tuned amplifiers

In eqn. (5.7) the impedance of a parallel tuned circuit was given as

$$Z = \frac{Rd}{d + j[(f/f_0) - (f_0/f)]}.$$

Since

$$\frac{f}{f_0} - \frac{f_0}{f} \doteq 2\frac{(f - f_0)}{f_0}, \quad \text{and} \quad d = \frac{B}{f_0},$$

$$Z = R \cdot \frac{B/2}{B/2 + jx},$$

where $x = f - f_0$, the frequency off resonance. Then making $B = 2$ gives

$$\frac{Z}{R} = \frac{1}{1 + jx}, \tag{5.22}$$

$$\left|\frac{Z}{R}\right| = \frac{1}{\sqrt{(1 + x^2)}} \tag{5.23}$$

This is the equation of the normalized selectivity curve of Fig. 5.9 and from it the frequency response of any parallel-tuned circuit may be obtained.

Equation (5.23) may be written as $1/\sqrt{(1 + x^{2n})}$, so that the normalized selectivity curve represents the case of $n = 1$. The way in which the value of n modifies the shape of the response is shown

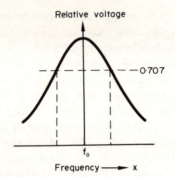

Fig. 5.9. Normalized selectivity curve $\left(\left|\dfrac{Z}{R}\right| = \dfrac{1}{\sqrt{(1+x^2)}}\right)$.

in Fig. 5.10. As n increases, the 3-dB bandwidth remains constant but the response flattens, and the cut-off outside the passband becomes sharper.

By means of Butterworth's technique[36] it can be shown that each of these responses may be synthesized by a number of staggered tuned stages as indicated in Table 5.1, where b is the bandwidth of a single stage.

It will be noticed that most single-stage bandwidths are less than the over-all bandwidth. A better over-all gain–bandwidth product is therefore obtainable than with cascaded synchronous circuits. It is for this reason, together with the sharper cut-off outside the passband, that staggered circuits are used.

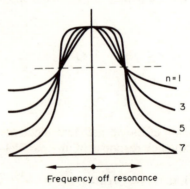

Fig. 5.10. $\left|\dfrac{Z}{R}\right| = \dfrac{1}{\sqrt{(1+x^{2n})}}$ plotted for various values of n.

Table 5.1. Staggered-tuned stages to provided bandwidth B at frequency f_0.

n	Single-tuned stages required
2	Two stages at $f_0 \pm 0.35B$, each having $b = 0.71B$.
3	Two stages at $f_0 \pm 0.43B$, each having $b = 0.5B$.
	One stage at f_0 and $b = B$.
4	Two stages at $f_0 \pm 0.46B$, each having $b = 0.38B$.
	Two stages at $f_0 \pm 0.19B$, each having $b = 0.92B$.
5	Two stages at $f_0 \pm 0.48B$, each having $b = 0.31B$.
	Two stages at $f_0 \pm 0.29B$, each having $b = 0.81B$.
	One stage at f_0 and $b = B$.

DESIGN EXAMPLE 5.3

Required, an amplifier having a response of the form $1/\sqrt{(1 + x^6)}$, ($n = 3$), at a frequency of 50 MHz and bandwidth of 6 MHz. Over-all gain to be 60 dB.

Since $n = 3$, from Table 5.1, there is required:

(a) a stage at 50 MHz with $b = 6$ MHz;
(b) a stage at $f_0 + 0.43B = 52.78$ MHz with $b = 3$ MHz;
(c) a stage at $f_0 - 0.43B = 47.22$ MHz with $b = 3$ MHz.

Gain per stage $= 20$ dB $= 10$.

The gain–bandwidth products of the three stages are therefore 60 MHz for (a) and 30 MHz for (b) and (c).

From eqn. (5.10) gain–bandwidth product $= g_m/2\pi C$, which is greatest for stage (a). Hence, allowing 28 pF as before, the required g_m is

$$g_m = 60 \times 10^6 \times 2\pi \times 28 \times 10^{-12} = 10.5 \text{ mS}.$$

An active device having a g_m of say 12 mS would be suitable and the detailed design of each stage is carried out as in Design Example 5.1.

5.6. Double-tuned circuits

Double-tuned circuits are employed where a passband is required centred on a fixed frequency, and coupling between the two circuits may be inductive, capacitive or a combination of both. The most common type is that of inductive coupling as is used, for instance, in the intermediate frequency section of a superhet radio receiver. Only this method of coupling is therefore considered.

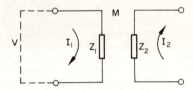

Fig. 5.11. Equivalent networks of an inductively coupled circuit.

Inductively coupled circuits

Inductively coupled circuits can be represented by the equivalent circuit of Fig. 5.11, in which Z_1 is the primary impedance, Z_2 is the secondary impedance and $M = k\sqrt{(L_1L_2)}$ is the mutual inductance that exists between them. The performance of the circuit may be examined as follows:

Impedance coupled from secondary into
primary $= (\omega M)^2/Z_2$.   (5.24)

Equivalent primary impedance $= Z_1 + (\omega M)^2/Z_2$.   (5.25)

Primary current $I_1 = \dfrac{V}{Z_1 + (\omega M)^2/Z_2}$.   (5.26)

Voltage induced in secondary $= -j\omega M I_1$.   (5.27)

Secondary current $I_2 = \dfrac{-j\omega M I_1}{Z_2} = \dfrac{-j\omega M V}{Z_1 Z_2 + (\omega M)^2}$.   (5.28)

When the mutual inductance is small, and the secondary impedance is large, the coupled impedance is small. Under these conditions the primary current is almost the same as if no secondary were present. If, however, Z_2 is small and M is not small, then the coupled impedance is significant. When Z_2 is reactive with a given phase angle, the coupled impedance has the same phase angle but with the sign reversed. When Z_2 is purely resistive the coupled impedance is also resistive.

Two resonant circuits, inductively coupled

The behaviour of a pair of inductively coupled circuits, resonant at the same frequency, is largely determined by the degree of coupling between them. In Fig. 5.12 is shown the way primary and

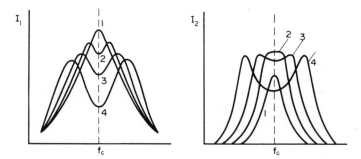

Fig. 5.12. Primary and secondary responses of inductively coupled tuned circuits for various coefficients of coupling.

secondary currents vary with frequency for four values of k, the coefficient of coupling.

Curve 1 represents a very small coefficient of coupling and both primary and secondary responses have approximately the same shape as would be obtained from single-tuned circuits. If the coupling is increased until the coupled resistance, at resonance, is equal to the primary resistance, the response of curve 2 results. Such a coupling is known as the *critical* coefficient of coupling, and provides the maximum value of secondary current.

Thus, for critical coupling,

$$(\omega M)^2/R_2 = R_1. \qquad (5.29)$$

Substituting $M = k\sqrt{(L_1 L_2)}$ and rearranging gives

$$\text{Critical coefficient of coupling } k_c = 1/\sqrt{(Q_1 Q_2)}. \qquad (5.30)$$

If the coupling is increased beyond the critical value, as in curves 3 and 4, two humps appear in each response, and these become more pronounced and more widely spaced, the tighter the coupling is made. This is explained as follows. The secondary impedance is $R_2 + j[\omega L - (1/\omega C)]$, and the coupled impedance is $(\omega M)^2/\{R_2 + j[\omega L - (1/\omega C)]\}$. At resonance this is a maximum and is resistive and, since it is effectively in series with the primary impedance, the primary current at resonance is a minimum. At frequencies above and below resonance the secondary impedance is reactive, and the resulting coupled impedance is also reactive but of reversed sign. For instance, an inductive reactance is coupled into the primary as a capacitive reactance. This neutralizes some of the primary reactance, lowering the primary impedance and increasing the primary current. Circuits which display this double-humped effect are said to be over-coupled.

If the two tuned circuits have equal Q's, the magnitude of the humps are equal. If, however, Q_1 does not equal Q_2 and if the humps are widely spaced, the low-frequency hump will tend to be the greater of the two.

Design considerations

An equivalent circuit is given in Fig. 5.13 in which R_1 and R_2 represent the loading of the tuned circuits, including the r_d of FET 1 and the input resistance of FET 2. The gain of the stage will be

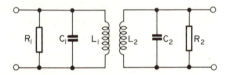

Fig. 5.13. Equivalent network of two inductively coupled tuned circuits, with the effects of loading represented by R_1 and R_2.

largely determined by the dynamic resistance of the tuned circuits and hence, for high gain, L should be large compared with C. There is a limit, however, to the L/C ratio which may be used, because for frequency stability reasons stray capacitance should not form a large proportion of the total tuned circuit capacitance.

Since the circuit is to operate at a fixed centre frequency, it is important that no detuning should take place due to temperature changes. Permeability tuning is therefore preferred, as this is more stable than capacitor tuning, and the tuned circuit capacitors are usually of the silvered mica type.

The development of design equations for the bandpass circuit is somewhat laborious, particularly for the general case of unequal Q's, and so is not given here. For the special, and usual case of equal high Q's with critical coupling, the relevant equations are as follows:

3 dB bandwidth of a single stage

$$B = \frac{f_0\sqrt{2}}{Q} = \frac{\sqrt{2}}{2\pi RC}. \tag{5.31}$$

Single-stage gain–bandwidth product,

$$GB = \frac{\sqrt{(2)}g_m}{4\pi C}, \tag{5.32}$$

where

$$C = \sqrt{(C_1 C_2)} \quad \text{and} \quad R = \sqrt{(R_1 R_2)}.$$

Critical coefficient of coupling

$$k_c = 1/Q, \tag{5.33}$$

and combining eqns. (5.31) and (5.33),

$$k_c = \frac{1}{2\pi f_0 RC} = \frac{B}{f_0\sqrt{2}}. \tag{5.34}$$

When n circuits of this type are cascaded the over-all bandwidth is less than a single-stage bandwidth in the ratio

$$\frac{\text{Over-all bandwidth}}{\text{Single-stage bandwidth}} = \frac{1}{1.1\, n^{1/4}}. \tag{5.35}$$

This expression is very accurate for a large n and is within 10 per cent of the correct value when $n = 2$.[36]

Design steps

1. Estimate the stray capacitance associated with each tuned circuit and select a value for each fixed capacitor which will make $C = C_1 = C_2$.
2. Determine the value of inductance which will resonate with C at the centre frequency.
3. Calculate the gain–bandwidth product of a single stage. If more than one stage is required to obtain the specified gain, use eqn. (5.35) to determine what bandwidth is necessary in each stage.
4. Substitute for B in eqn. (5.31) thus obtaining the Q for each circuit, and from eqn. (5.33) determine the critical coefficient of coupling.

5. Consider the effect of transistor loading on the primary and secondary circuits, and evaluate the damping resistors if these are required.

DESIGN EXAMPLE 5.4 (Fig. 5.14)

Required, a bandpass amplifier having a centre frequency of 10 MHz and a bandwidth of 250 kHz, gain to be at least 2500.

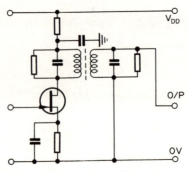

Fig. 5.14. Completed circuit of Design Example 5.4.

For a good gain–bandwidth product an FET with a high g_m is necessary. Let the device chosen be a BF 246. From the data sheets for this transistor, if $V_{DS} = 15$ V and $V_{GS} = -3$ V, $I_D = 12$ mA and $g_m = 16$ mS.

Let $C = C_1 = C_2 = 100$ pF (including stray capacitance). Single-stage gain–bandwidth product

$$= \frac{\sqrt{(2)}g_m}{4\pi C} = \frac{1.41 \times 16 \times 10^{-3}}{12.56 \times 100 \times 10^{-12}} \doteqdot 18 \text{ MHz}.$$

Therefore

$$\text{gain} = \frac{18 \times 10^6}{250 \times 10^3} = 72.$$

A gain of 2500 is thus readily obtainable with two stages. From eqn. (5.35), for an over-all bandwidth of 250 kHz, single-stage bandwidth $(n = 2) = 1.1 \times 2^{1/4} \times 250 \times 10^3 = 325$ kHz.

Checking gain with this new bandwidth, $GB = 18$ MHz, so gain per stage = 18 MHz/325 kHz = 55. Sufficient gain is still therefore available.

$$L = \frac{1}{4\pi^2 f_0^2 C} = \frac{1}{40 \times 10^{14} \times 10^{-10}} = 2.5 \ \mu\text{H}.$$

From eqn. (5.31) the required

$$Q = \frac{f_0\sqrt{2}}{B} = \frac{10 \times 10^6 \times 1.41}{325 \times 10^3} = 43,$$

and $k_c = 1/Q = 0.023$.

The primary tuned circuit is loaded by the r_d of T_1, which is $500 \text{ k}\Omega$. Similarly the secondary of the first stage is loaded by the input resistance of T_2. However, at 10 MHz this is sufficiently high to be ignored.

From eqn. (5.31), the value of the effective damping resistance should be

$$R_{\text{eff}} = \frac{\sqrt{2}}{2\pi BC} = \frac{1.41}{6.28 \times 325 \times 10^3 \times 100 \times 10^{-12}} = 6.9 \text{ k}\Omega.$$

This must be the equivalent value of the damping resistor in parallel with the dynamic resistance of the tuned circuit and the FET loading resistance.

Assume that the transformer has primary and secondary Q's of 100. Then,

$$R_D = Q\omega_0 L = 100 \times 6.28 \times 10^7 \times 2.5 \times 10^{-6} = 15.7 \text{ k}\Omega.$$

To achieve required R_{eff} for the secondary, the damping resistor R must be

$$R = \frac{R_D \cdot R_{\text{eff}}}{R_D - R_{\text{eff}}} = \frac{15.7 \text{ k}\Omega \times 6.9 \text{ k}\Omega}{15.7 \text{ k}\Omega - 6.9 \text{ k}\Omega} = 12.3 \text{ k}\Omega.$$

The dynamic resistance of the primary winding is shunted by $r_d = 500 \text{ k}\Omega$ giving an equivalent shunt resistance of $R_s = 15.2 \text{ k}\Omega$, and the required primary damping resistor

$$R = \frac{R_s \cdot R_{\text{eff}}}{R_s - R_{\text{eff}}} = \frac{15.2 \text{ k}\Omega \cdot 6.9 \text{ k}\Omega}{8.3 \text{ k}\Omega} = 12.6 \text{ k}\Omega.$$

Both damping resistors may therefore be made $12 \text{ k}\Omega$.

Decoupling. The design values for r_d and g_m were selected for $V_D = 15 \text{ V}$. Assuming an available supply of 20 V, the extra 5 V may be dropped in a decoupling resistor. Thus, $R = V/I = 5/(12 \times 10^{-3}) \doteq 400 \ \Omega$.

Let it be the preferred value of 470 Ω. The decoupling capacitor should offer a signal path to earth which is small compared to this. At 10 MHz a 330 pF capacitor has a reactance of 48 Ω which is suitable.

The correct bias conditions are set up as in Design Example 5.1. In this example, the problem of unequal loading of the two tuned circuits has been easily overcome because the specification is not particularly stringent. In cases where the required Q's are high, and where the relative values of r_d and r_g are such as to give an unbalanced condition, a technique described by Langford-Smith[38] may be used. In this treatment a transformer is designed having equal values of primary and secondary Q when unloaded. Designating them by Q_{u1} and Q_{u2} they should have such values that, when in circuit and loaded by r_d and r_g, $Q = \sqrt{(Q_{u1} \cdot Q_{u2})}$, where Q is that required to provide the correct gain and bandwidth.

5.7.   Tuned amplifiers using bipolar transistors

The necessary theory associated with parallel-tuned circuits has been presented in § § 5.1 and 5.6. Before proceeding with the design of a tuned amplifier, the characteristics of bipolar transistors, when used at high frequencies, must first be studied.

Choice of circuit configuration

The frequency response of a transistor may be expressed in terms of the frequency at which the current gain is reduced to $1/\sqrt{2}$ of its low-frequency value. At this frequency the phase shift starts to change rapidly, and this can be of importance in obtaining stable amplification over a range of frequencies. The frequency at which this occurs is called the cut-off frequency and is denoted by f_α for the common base and f_β for the common emitter configuration. To a rough approximation $f_\alpha = \beta f_\beta$, i.e. a higher cut-off frequency is obtainable with common base than with common emitter. Despite this fact, the common base arrangement is not necessarily preferred at high frequencies. Transistors are available with sufficiently high values of f_β that they may be used, in IF amplifiers for instance, in the common emitter configuration, taking advantage of the higher current gain of that circuit arrangement.

High-frequency equivalent network

At high frequencies the equivalent networks of Fig. 1.34 fail to yield accurate design results because of the h.f. effects which are inherent in the structure of a transistor. To take account of these effects the modified-T and hybrid-π networks provide the most convenient representations for common base and common emitter respectively. Both of these are shown in Fig. 5.15, but attention here will be limited to the hybrid-π network.

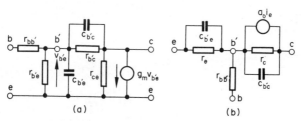

Fig. 5.15. The hybrid-π and modified-T equivalent networks for use at high frequency. The symbol b' represents the internal base point.

The following relationships hold between the low frequency h parameters of Fig. 1.34 and the hybrid-π parameters:

$$r'_{be} = h_{fe}/g_m, \qquad r'_{bb} = h_{ie} - r'_{be},$$

$$r'_{bc} = r'_{be}/h_{re}, \qquad r_{ce} = 1/[h_{oe} - (1 + h_{fe})g'_{bc}],$$

and $\qquad c'_{b'e} = \beta/(2\pi f_1 r'_{be}).$

The symbols used have the following significance:

$r_{b'b}$ represents the ohmic resistance which connects the active part of the base to the external base connector b.

$r_{b'e}$ takes account of recombination base current resulting from an increase of minority carriers in the base region.

$r_{b'c}$ represents the effect of feedback between c and b'.

$c_{b'e}$ is the emitter diffusion capacitance.

$c_{b'c}$ is the collector capacitance.

f_1 is the frequency at which the modulus of the common emitter forward current transfer ratio (h_{fe}) is unity, and

β is the current gain for common emitter connection.

Since r'_{be} is usually small, the effect of c'_{be} is frequently negligible. The collector capacitance, however, is more significant since it shunts a medium to high resistance. This capacitance can seriously reduce bandwidth, and also provides an internal feedback path from the collector output to the input. With good h.f. transistors r'_{be} and r_{ce} will often be so large that they can be ignored. Ignoring r_{ce} implies a load impedance which is small compared with it, while r'_{bc} may be ignored because at h.f. it is large compared with the reactance of c'_{bc}.

The hybrid-π of Fig. 5.15a may therefore often be reduced to the simplified network of Fig. 5.16.

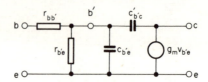

Fig. 5.16. Simplified hybrid-π network.

5.8.   Neutralization[39]

As frequency increases, the reactance of c'_{bc} falls, thus increasing feedback from output of the transistor. In most instances it is necessary to cancel the effects of this internal feedback. The term unilateralization describes the process by which complete cancellation is obtained. A more realistic approach is that of neutralization in which only partial cancellation is achieved, but this being sufficient to enable the effects of feedback to be ignored.

In Fig. 5.17 is shown a hybrid-π equivalent circuit with a feedback network Z_n fed with a signal derived from the output voltage. For the cancellation of feedback effects, v_1 should be zero for any value of v_2. Employing superposition, v_1 may be written as the sum of two terms, one from v_2 and the other from the Av_2 generator. The circuit may be redrawn as in Fig. 5.18, in which

$$Z_1 = \frac{r'_{be}}{1 + sc'_{be}r'_{be}},$$

$$v'_2 \text{ (due to } v_2) = v_2 \cdot \frac{Z_1}{Z_1 + 1/sc'_{bc}}.$$

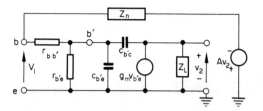

Fig. 5.17. Hybrid-π network with external feedback.

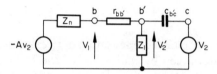

Fig. 5.18. Modification of Fig. 5.17 for the application of the superposition theorem.

Making use of Thevenin's theorem the circuit simplifies further to that of Fig. 5.19, in which

$$Z_2 = \frac{Z_1}{1 + sc_{b'c}Z_1}.$$

Using superposition,

$$v_1 \text{ (due to } v_2) = \left(\frac{Z_n}{Z_n + r_{b'b} + Z_2}\right)\left(\frac{Z_1}{Z_1 + 1/sc_{b'c}}\right)v_2,$$

$$v_1 \text{ (due to } Av_2) = -A\left(\frac{Z_2 + R_{b'b}}{Z_n + r_{b'b} + Z_2}\right)v_2,$$

therefore,

$$v_1 = \left(\frac{Z_n}{Z_n + r_{b'b} + Z_2}\right)\left(\frac{Z_1}{Z_1 + 1/sc_{b'c}}\right)v_2$$

$$-A\left(\frac{Z_2 + r_{b'b}}{Z_n + r_{b'b} + Z_2}\right)v_2,$$

For $v_1 = 0$,

$$\frac{Z_n Z_1}{Z_1 + 1/sc_{b'c}} = A(Z_2 + r_{b'b}).$$

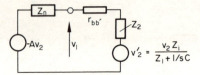

Fig. 5.19. The equivalent network further simplified by the use of Thevenin's theorem.

Solving for Z_n and substituting for Z_1 and Z_2 yields

$$Z_n = A r'_{bb} \left(1 + \frac{C_{b'e}}{C_{b'c}}\right) + \frac{A}{s C'_{bc}} \left(1 + \frac{r_{b'b}}{r_{b'e}}\right). \qquad (5.36)$$

The required feedback compensation network is therefore a series R–C circuit determined by the values of the transistor resistances and capacitances and by the feedback ratio A. For the capacitor, it is often a reasonable approximation to make its value c'_{bc}/A. In many practical circuits, where the transistor is used at frequencies below f_β, the resistor is omitted and the capacitor alone provides adequate neutralization.

Choice of transistor

The neutralizing network shunts both input and output impedance of the stage and its capacitive component forms part of the tuning capacitance. A transistor should therefore be chosen having a small value of c'_{bc} and thus requiring a small feedback capacitor. It is also desirable that the component parts of the internal feedback path should not be subject to a very large spread. For operation in common emitter mode, the transistor should have a common-base cut-off frequency approximately 10 times the frequency at which the stage is to be operated.

Design considerations

In the case of the FET bandpass amplifier, the design centred on a voltage amplifier having large input and output impedances, and the efficiency of power transference was largely ignored. With bipolar transistors the input and output impedances are comparatively small and hence the effect on the Q of the tuned circuit is greater. Since these impedances are considerably different from each other, it is necessary that, in addition to providing the necessary bandwidth and selectivity, the tuned transformer should also act as an impedance matching device in order to achieve an efficient transference of power.

The circuit takes the form of Fig. 5.20, and the AV_2 term is obtained from the transformer secondary, A being the transformation ratio. The collector is connected to a tap on the primary to

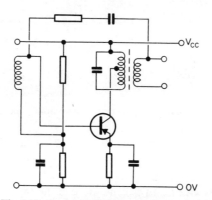

Fig. 5.20. Basic circuit of Design Example 5.5.

reduce the loading effect which would otherwise occur if the transistor output impedance was across the whole primary. This also enables the tuning capacitor to be of a reasonable value.

For maximum transference of power, the transformer should have a turns ratio,

$$n = \sqrt{\left(\frac{R_{out}}{R_{in}}\right)}, \tag{5.37}$$

where R_{out} is the output resistance and R_{in} is the input resistance of the next stage. To determine the value of the input and output impedances, use is made of the fact that, due to the neutralizing network, v_2 has no effect on v_1 and vice versa. It is therefore permissible to short circuit v_2, and the input impedance is then formed by the circuit of Fig. 5.21. This may be resolved into a resistance R_{in} and a capacitance C_{in} in parallel. Similarly, by short circuiting v_1, the output impedance may be calculated, and the equivalent circuit redrawn as in Fig. 5.22.

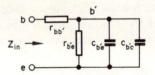

Fig. 5.21. Equivalent network of the input impedance of the neutralized i.f. amplifier stage.

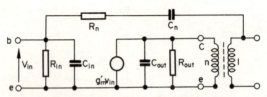

Fig. 5.22. Equivalent network to represent the output impedance of the stage.

This circuit may be further simplified by considering the shunting effects of the neutralizing circuit and then treating the input and output as two independent circuits. Thus, R_n and C_n may be resolved into R_p and C_p in parallel across the input circuit, and also across the secondary of the output transformer. It appears as $n^2 R_p$ and C_p/n^2 across the transistor output impedance and obviously has less effect there than it does across the input circuit. The final equivalent circuit is therefore of the form of Fig. 5.23.

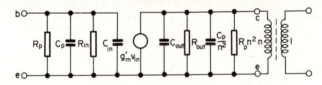

Fig. 5.23. Final equivalent network of the bandpass amplifier of Fig. 5.20.

Design steps

1. Decide under what d.c. conditions the transistor is to be operated, and calculate the values of emitter and base resistors and bypass capacitors in the usual way.
2. Draw a hybrid-π equivalent circuit and calculate the value of R_{out} and R_{in}, and thus determine the turns ratio n.

3. From eqn. (5.36) calculate R_n and C_n for the neutralizing circuit.
4. Calculate the equivalent impedance which represents the shunting effect of the neutralizing components across the input and output and construct the final equivalent circuit.
5. With the knowledge of the bandwidth required from each stage, use expression (5.9) to determine the tuned circuit effective Q.
6. Assume a value for the unloaded Q, select the tuned circuit capacitor C, calculate L to resonate with C at the centre frequency, and then determine the values of r, and L/Cr, the equivalent shunt resistor of the unloaded circuit.
7. Calculate the required shunting resistance across the tuned circuit to provide the correct bandwidth and thus establish what resistance should be reflected from the transistor output to achieve this.
8. Knowing this value, calculate at what point the transformer primary should be tapped.

DESIGN EXAMPLE 5.5

Required, a transistor i.f. amplifier to operate at a centre frequency of 470 kHz.

From the makers' data sheets, a 2N 1302 has a typical α cut-off frequency of 12 MHz when operated at a collector voltage of 6 V and with an emitter current of 1 mA. This is more than 10 times the required centre frequency so this transistor is suitable. The circuit arrangement will be that of Fig. 5.20.

Setting up the d.c. conditions. Let R_E be 820 Ω, so that $V_E = 0.82$ V, with an emitter current of 1 mA. For a V_C of 6 V, and allowing for V_E and the voltage drop across the tuned circuit, $V_{CC} = 7$ V. Thus making $V_B = 1$ V, $7R_2/(R_1 + R_2) = 1$ and $6R_2 = R_1$.

Also, make $R_1 R_2/(R_1 + R_2) = 10 R_E = 8.2$ kΩ.

Therefore,

$$R_1 R_2 = 8.2 \, kR_1 + 8.2 \, kR_2.$$

$$6R_2^2 = 49.2 \, kR_2 + 8.2 \, kR_2 \quad \text{and} \quad 6R_2 = 57.4 \, k\Omega.$$

Let R_2 be 10 kΩ and hence R_1 should be 56 kΩ. Suitable decoupling capacitors C_E and C_1 are 0.25 μF and 0.1 μF respectively.

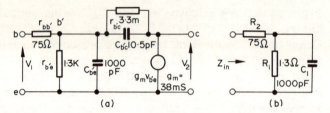

Fig. 5.24. Equivalent hybrid-π network of the circuit of Design Example 5.5 with an equivalent representation of the input impedance if v_2 is short circuited.

Input and output impedances. Based on Fig. 5.15a, and using values from the makers' data sheets, the hybrid-π equivalent network is drawn as in Fig. 5.24a. In Fig. 5.24b is drawn the effective input circuit if v_2 is short circuited, and if c'_{bc} and r'_{bc} are ignored.

$$Z(R_1, C_1) = \frac{jR_1X_1}{R_1 + jX_1} = \frac{R_1X_1^2}{R_1^2 + X_1^2} + \frac{jR_1^2X_1}{R_1^2 + X_1^2}.$$

Substituting values this becomes a resistance of 78 Ω in series with a capacitive reactance of 310 Ω. Adding $r'_{bb} = 75\ \Omega$, $R_s = 153\ \Omega$ and $X_s = 310\ \Omega$.

This is now converted into an equivalent parallel circuit of R_p and C_p.

Admittance

$$Y = G + jB = \frac{R_s}{R_s^2 + X_s^2} + \frac{jX_s}{R_s^2 + X_s^2},$$

therefore,

$$R_p = \frac{R_s^2 + X_s^2}{R_s} \quad \text{and} \quad X_p = \frac{R_s^2 + X_s^2}{X_s}.$$

From which $R_p = R_{in} = 780\ \Omega$ and $X_p = 386\ \Omega$, so $C_p = C_{in} = 875$ pF.

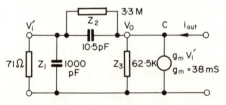

Fig. 5.25. Equivalent network of the output circuit, with v_1 of Fig. 5.24 short circuited.

By short circuiting v_1 in Fig. 5.24a the output circuit becomes that of Fig. 5.25.

Making use of nodal analysis as discussed in Appendix A, at node b',

$$\frac{v_1'}{Z_1} + \frac{v_1' - v_0}{Z_2} = 0; \qquad (5.38)$$

at node C,

$$\frac{v_0}{Z_3} + \frac{v_0 - v_1}{Z_2} + g_m v_1' = i_0. \qquad (5.39)$$

From eqn. (5.38)

$$v_1' = Z_1 v_0 / (Z_1 + Z_2).$$

Substituting in eqn. (5.39) and rearranging,

$$v_0 \left[\frac{1}{Z_3} + \frac{1}{Z_2} + \left(\frac{Z_1}{Z_1 + Z_2} \right) \left(g_m - \frac{1}{Z_2} \right) \right] = i_0.$$

Therefore output impedance,

$$Z_0 = \frac{v_0}{i_0} = \frac{1}{(1/Z_3) + (1/Z_2) + [Z_1(Z_1 + Z_2)][g_m - (1/Z_2)]}.$$

Calculating the component parts of this expression separately yields:

$$Z_1 = (67.9 - j14.28); \quad 1/Z_1 = (14 + j2.96)10^{-3},$$
$$Z_2 = (0.31 - j31.8)10^3; \quad 1/Z_2 = (0.3 + j31)10^{-6},$$
$$Z_3 = 62.5 \times 10^3; \quad 1/Z_3 = 16 \times 10^{-6}.$$

Thus,

$$1/Z_3 + 1/Z_2 = (16.3 + j31)10^{-6},$$
$$Z_1/(Z_1 + Z_2) = (0.474 + j2.128)10^{-3},$$
$$g_m - 1/Z_2 = (38{,}000 - j31)10^{-6}.$$

Substituting these values in eqn. (5.40),

$$Z_0 = \frac{(34.4 - j111.9)10^6}{13{,}705},$$

i.e. a resistance $R_s = 2.51 \text{ k}\Omega$ in series with a rectance $X_s = 8.16 \text{ k}\Omega$.

Resolving into equivalent parallel components,

$$R_{out} = \frac{R^2 + X^2}{R} = 29\,k\Omega,$$

$$X_p = \frac{R^2 + X^2}{X} = 8.16\,k\Omega \quad \text{so } C_{out} = 38\,pF.$$

Transformer turns ratio

$$n = \sqrt{\left(\frac{29,000}{780}\right)} \doteqdot 6.$$

Neutralizing circuit

$$R_n = A r'_{bb}\left(1 + \frac{c'_{be}}{c'_{bc}}\right) = 75/6\left(1 + \frac{1000}{10.5}\right) = 1.188\,k\Omega.$$
$$C_n = c'_{bc}/A = 6 \times 10.5 = 63\,pF.$$

Let them be the preferred values of 1.2 kΩ and 68 pF.

Shunting effect on input. At 470 kHz the reactance of 68 pF $\doteqdot$ 5 kΩ.

$$R_{pi} = \frac{R_n^2 + X_n^2}{R_n} = \frac{(1.44 + 25)10^6}{1.2 \times 10^3} = 22\,k\Omega,$$

$$X_{pi} = \frac{R_n^2 + X_n^2}{X_n} = \frac{26.44 \times 10^6}{5 \times 10^3} = 5.28\,k\Omega.$$

Thus, $C_{pi} = 6\,pF$.

Shunting effect on output.

$$R_{po} = n^2 R_{pi} = 792\,k\Omega,$$
$$C_{po} = C_{pi}/n^2 = 1.8\,pF.$$

Assuming that the stage is followed by a similar stage having the same input impedance, the final equivalent circuit is that of Fig. 5.26. In this the conductance of the representative current generator has been given a value of 35 instead of 38 mS. This is necessary because the current is now shown as a function of v_{in}, at the base connection, and not of v'_{be}. The difference in g_m takes account of the voltage lost in r'_{bb}.

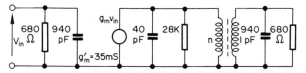

Fig. 5.26. Final equivalent circuit of Design Example 5.5. Note the different value of g'_m to take account of the voltage drop in $r_{b'b}$.

The capacitance reflected from the secondary into the primary of Fig. 5.25 is $940/n^2 = 26\,\text{pF}$, giving a total primary capacitance of 66 pF. Similarly, the resistance reflected from the secondary into the primary is $680n^2 = 24.5\,\text{k}\Omega$, giving a total shunt resistance of about 13 kΩ. To reduce the loading effect of this resistance on the tuned circuit, a tapped primary winding is used.

The tuned primary. Assuming that the complete i.f. amplifier is to consist of two stages, and that the over-all bandwidth is to be 7 kHz, then each stage must have a bandwidth of $7 \times 1.2\sqrt{2} = 9\,\text{kHz}$.

From eqn. (5.9), bandwidth $= f_0/Q = 1/2\pi RC$, from which $Q_{\text{eff}} = 470/9 = 52$, and equivalent shunt resistance $R_{\text{eq}} = 70\,\text{k}\Omega$.

In circuits of this type, a transformer would be used having an unloaded primary Q_u of 100, and tuned by a 250-pF capacitor. To resonate with this capacitor at 470 kHz, if the reflected capacitance from the transistor output is assumed to be negligible, $L = 450\,\mu\text{H}$.

Primary resistance $r = \omega L/Q_u = 13.3\,\Omega$.

Equivalent shunt resistance $R_D = L/Cr \doteqdot 135\,\text{k}\Omega$.

It is therefore necessary that the resistance reflected from the transistor output across the primary should reduce the equivalent

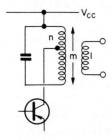

Fig. 5.27. Circuit to illustrate the auto-transformer action of the primary winding.

shunt resistance from 135 kΩ to the required value of 70 kΩ. The reflected resistance should thus appear as 145 kΩ and this is achieved by the auto-transformer action of the tapped primary winding. Referring to Fig. 5.27, the output resistance of 13 kΩ across n appears across the whole primary m as 13 k$\Omega \times m^2/n^2$. For this to equal 145 kΩ, $m^2/n^2 = 11$ so that $m/n = 3.3$. For matching purposes it has already been decided that $n : 1 = 6$, so the complete primary to secondary turns ratio becomes approximately 20:1. The voltage gain of the circuit, measured from transistor input to transformer secondary output, is $g_m R/m$, thus,

$$\text{Gain} = 35 \times 70 \times 1/20 = 122.$$

The completed circuit of the first stage is given in Fig. 5.28. The second stage of the amplifier, which would normally be feeding into a detector circuit, would be modified to take account of the different input impedance of that circuit.

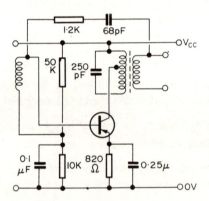

Fig. 5.28. Completed circuit of Design Example 5.5.

5.9   Integrated circuits

The design of a discrete component tuned amplifier can be divided into two parts. There is firstly the active device, the transistor, which amplifies an incoming signal and provides the necessary drive to a resonant circuit. Here the task is to select a suitable transistor and to arrange that it is correctly biased. Secondly, L–C circuits are designed to resonate at the desired centre frequency. In the design of these, the selected capacitance will include the input and output capacitance of the transistor, and other stray capacitances. Additionally the input and output resistances of the active devices are considered, since they load the tuned circuits, when the bandwidth is computed.

The radio and television industry represents a consumer field which, in magnitude, is comparable with the computing field, and is of similar attraction to the integrated circuit manufacturers. Accordingly, development has been aimed at the application of linear amplifiers to this industry, and the first widely used device was the Fairchild μA703 direct-coupled amplifier, introduced in 1967 for use in r.f. and i.f. stages. The eight lead package in which it is mounted includes the necessary biasing network, and its low internal feedback provides greater stability than is available from comparable discrete component circuits. It can additionally be used as a mixer or oscillator at frequencies up to 150 MHz. From the makers' data sheets, the amplifier has a typical g_m of 33 mS. Its input impedance is comprised of a resistance of 3.5 kΩ and capacitance of 10 pF, and its output impedance 30 kΩ resistance and 2 pF capacitance. In the design of an amplifying stage using these devices, the task is thus reduced to the selection of components for the necessary tuned circuits and decoupling networks, according to the principles outlined in previous sections.

Following the μA 703, there has been introduced a range of integrated circuits to serve as audio amplifiers, stereo multiplex decoders, audio pre-amplifiers, and complete radios, etc. In each case the device replaces the discrete active components and their biasing circuits, and the design task is the provision of the necessary external circuitry.

APPENDIX A

Solutions of Simple Network Problems

BASIC NETWORK THEOREMS[60,61]

(a) Kirchhoff's Current Law (KCL)

The algebraic sum of currents flowing into a point (or node) is zero. Thus, in Fig. A.1,

$$i_1 + i_2 + i_3 = 0. \tag{A.1}$$

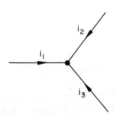

Fig. A.1. Kirchhoff's current law.

(b) Kirchhoff's Voltage Law (KVL)

The sum of voltages around a closed loop is zero. Thus in Fig. A.2,

$$v_1 + v_2 + v_3 = 0. \tag{A.2}$$

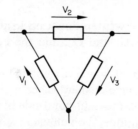

Fig. A.2. Kirchhoff's voltage law. The arrows indicate that the instantaneous voltage is measured from head to end.

xvii

EXAMPLE A.1

It is required to find the voltage across R_3 of Fig. A.3.

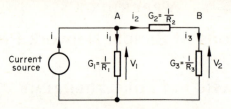

Fig. A.3. Network for example A.1. The elements are drawn as conductances for nodal analysis.

By KCL, the current flowing into node A is equal to the current flowing out of it

$$i = i_1 + i_2, \tag{A.3}$$

and for node B,

$$i_2 = i_3. \tag{A.4}$$

From Ohm's Law,

$$i_1 = \frac{v_1}{R_1}, \quad i_2 = \frac{v_1 - v_2}{R_2}, \quad i_3 = \frac{v_2}{R_3}.$$

Using conductances in place of resistances, eqns. (A.3) and (A.4) can be written as follows:

$$(G_1 + G_2)v_1 - G_2v_2 = i \quad \text{(node } A), \tag{A.5}$$

$$-G_2v_1 + (G_2 + G_3)v_2 = 0 \quad \text{(node } B). \tag{A.6}$$

These equations can be written down on inspection. The elements associated with v_1 are G_1 and G_2, and the coupling element with node B is G_2.

Similarly, for node B, at which the voltage is v_2, the elements are G_2 and G_3, and the coupling element is again G_2. In this case there is no current generator and the right-hand side of the equation is zero.

From the above equations, the voltages v_1 and v_2 can be found as functions of the input current. A simple method of solving these simultaneous equations is as follows.

The coefficients of the left-hand sides of the equations are set out as an array,

$$\begin{vmatrix} (G_1 + G_2) & -G_2 \\ -G_2 & (G_2 + G_3) \end{vmatrix}.$$

If the elements of this system are cross-multiplied, the *determinant* of the equations is obtained, provided that the product of the upper right and lower left terms is made negative.

Thus, the determinant

$$\Delta = (G_1 + G_2)(G_2 + G_3) - (-G_2)^2,$$
$$= (G_1 G_2 + G_2 G_3 + G_3 G_1). \qquad (A.7)$$

Replacing the left-hand column of the coefficient array with the right-hand sides of eqns. (A.5) and (A.6) the following array is obtained:

$$\begin{vmatrix} i & -G_2 \\ 0 & (G_2 + G_3) \end{vmatrix}.$$

Cross-multiplying,

$$\Delta_{v1} = (G_2 + G_3)i. \qquad (A.8)$$

Similarly, if the right-hand side of the coefficient array is replaced by the right-hand sides of eqns. (A.5) and (A.6),

$$\begin{vmatrix} (G_1 + G_2) & i \\ -G_2 & 0 \end{vmatrix},$$

and the determinant is

$$\Delta_{v2} = G_2 i. \qquad (A.9)$$

Using eqns. (A.7), (A.8) and (A.9),

$$v_1 = \frac{\Delta_{v1}}{\Delta},$$
$$= \frac{(G_2 + G_3)i}{(G_1 G_2 + G_2 G_3 + G_3 G_1)}. \qquad (A.10)$$
$$v_2 = \frac{\Delta_{v2}}{\Delta},$$
$$= \frac{G_2 i}{(G_1 G_2 + G_2 G_3 + G_3 G_1)}. \qquad (A.11)$$

All the simultaneous equations in this book can be solved by this method which may be extended to deal with more complicated systems.[62] Since v_1 and v_2 are the voltages at the nodal points, eqns. (A.5) and (A.6) are the *nodal equations* and are formed by *nodal analysis*.

EXAMPLE A.2

Consider the network of Fig. A.4. If the dependent variables are assumed to be the loop or circuital currents i_1 and i_2, these can be

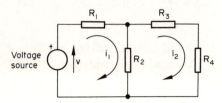

Fig. A.4. Network for example A.2. Loop currents i_1 and i_2 are drawn in a clockwise direction.

determined by the application of KVL. The currents are drawn flowing in a clockwise direction to assist in the formalization of the equations. Thus,

$$(R_1 + R_2)i_1 - R_2 i_2 = v, \quad \text{for loop 1,} \quad (A.12)$$

$$- R_2 i_1 + (R_2 + R_3 + R_4)i_2 = 0, \quad \text{for loop 2.} \quad (A.13)$$

Proceeding as in the previous example, the coefficient array is

$$\begin{vmatrix} (R_1 + R_2) & - R_2 \\ - R_2 & (R_2 + R_3 + R_4) \end{vmatrix}$$

and the determinant is

$$\Delta = (R_1 + R_2)(R_2 + R_3 + R_4) - (-R_2)^2,$$
$$= (R_1 R_2 + R_1 R_3 + R_1 R_4 + R_2 R_3 + R_2 R_4). \quad (A.14)$$

$$\begin{vmatrix} v & - R_2 \\ 0 & (R_2 + R_3 + R_4) \end{vmatrix}$$

so that

$$\Delta_{i1} = (R_2 + R_3 + R_4)v \quad (A.15)$$

and

$$i_1 = \frac{\Delta_{i1}}{\Delta} = \frac{(R_2 + R_3 + R_4)v}{R_1(R_2 + R_3 + R_4) + R_2(R_3 + R_4)}.$$

Also

$$\begin{vmatrix} (R_1 + R_2) & v \\ -R_2 & 0 \end{vmatrix},$$

so that

$$\Delta_{i2} = vR_2 \qquad\qquad (A.16)$$

and

$$i_2 = \frac{\Delta_{i2}}{\Delta} = \frac{vR_2}{R_1(R_2 + R_3 + R_4) + R_2(R_3 + R_4)}.$$

Equations (A.12) and (A.13) are *circuital equations* and are formed by *circuital analysis*.

APPENDIX B

Application of the Laplace Transform[62]

When a network contains reactive as well as resistive elements, the equations are in integral-differential form. For the LCR circuit shown in Fig. B.1, the equation resulting from the application of Kirchhoff's voltage law is

$$v = Ri + \frac{1}{C}\int_0^t i\,dt + L\frac{di}{dt}. \tag{B.1}$$

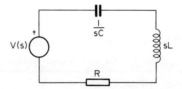

Fig. B.1. An L–C–R circuit with transformed impedances.

By using the Laplace Transformation it is possible to transform the equation into an algebraic form, which can then be treated in the same manner as the equations of Appendix A.

Equation (B.1) after transformation becomes

$$v(s) = Ri(s) + \frac{1}{sC}i(s) + sLi(s)$$

$$= \left(R + \frac{1}{sC} + sL\right)i(s), \tag{B.2}$$

where v and i are functions of the Laplace variable s instead of time as is implied in eqn. (B.1). Integration is indicated by an s in the denominator, and differentiation by an s in the numerator. Thus in Fig. B.1 R, C and L can be represented by the transformed

impedances R, $1/sC$ and sL. (If before the application of the voltage there is a charge on the capacitor or current in the inductance, the representation is a little more complex and reference should be made to a standard work on network theory.)[60]
Thus,

$$i(s) = \frac{v(s)}{R + (1/sC) + sL}.$$
(B.3)

If v is a sinusoidal signal, which is the case for a frequency response measurement, the Laplace variable s can be replaced by $j\omega$. The circuit current as a function of frequency is then

$$i(j\omega) = \frac{v(j\omega)}{R + (1/j\omega C) + j\omega L}.$$
(B.4)

If it is required to determine the form of the output for a given input signal, the equation can be changed back into the time domain by using the inverse transforms given in Table B.1. This method is employed in § 4.9 to find the time response of a capacitively coupled amplifier.

Table B.1 Laplace transform pairs

	Function of time	Function of s	
1.	$1(t > 0)$	$\dfrac{1}{s}$	Step function of unit amplitude applied at time $t = 0$
2.	$V(t > 0)$	$\dfrac{V}{s}$	Step function of amplitude V
3.	$\exp(-at)$	$\dfrac{1}{s+a}$ $\dfrac{1}{s}\dfrac{1}{1+(a/s)}$	
4.	$1 - \exp(-at)$	$\dfrac{a}{s(s+a)}$ $\dfrac{1}{s}\dfrac{1}{1+(s/a)}$	

EXAMPLE B.1

Figure B.2 represents the anode network of an amplifier. It is required to find the frequency response and the output signal

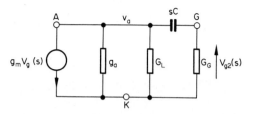

Fig. B.2. Transformed network for nodal analysis of the low-frequency performance of a valve amplifier.

resulting from the application of a rectangular step of voltage to the input terminals of the amplifier.

The nodal equations are written down by inspection, using the method given in Appendix A:

$$(g_a + G_L + sC)v_a(s) - sCv_{g2}(s) = -g_m v_{g1}(s), \qquad (B.5)$$

$$-sCv_a(s) + (G_G + sC)v_{g2}(s) = 0. \qquad (B.6)$$

$$v_{g2}(s) = \frac{-g_m sC v_{g1}(s)}{(g_a + G_L + sC)(G_G + sC) - (sC)^2}, \qquad (B.7)$$

$$= \frac{-g_m}{(G_G + g_a + G_L)} \cdot \frac{1}{1 + [G_G(g_a + G_L)/(G_G + g_a + G_L)sC]} v_{g1}(s),$$

or

$$\frac{v_{g2}(s)}{v_{g1}(s)} = A_0 \frac{1}{1 + (1/s\tau_2)}.$$

If the frequency response expression is required, the Laplace variable is replaced by $j\omega$.

Thus,

$$\frac{v_{g2}(j\omega)}{v_{g1}(j\omega)} = A_0 \frac{1}{1 + (1/j\omega\tau_2)}.$$

Since the Laplace transform of a rectangular step input of amplitude V is V/s (from Table B.1), the output as a function of s is

$$v_{g2}(s) = A_0 \frac{1}{1 + (1/s\tau_2)} \cdot \frac{V}{s}. \qquad (B.8)$$

By using transform pair 3 from Table B.1 the output response as a function of time can be obtained directly.

Thus,

$$v_{g2}(t) = A_0 \exp\left(-t/\tau_2\right)V. \tag{B.9}$$

The form of the response is shown graphically in Fig. 4.26.

APPENDIX C

Symbols used in this Book

British Standard Specification No. 3363, 1961, entitled "Letter symbols for light-current semi-conductor devices", makes recommendations for symbols to be used when describing transistor circuits. This system has been extended, in the same form, for use with FET and thermionic valve circuits and is stated briefly below.

Symbols

c or C,	Collector	
b or B,	Base	Bipolar transistor
e or E,	Emitter	
d or D,	Drain	
g or G,	Gate	Field effect transistor
s or S,	Source	
a or A,	Anode	
s or S,	Screen	
g or G,	Grid	Thermionic valve
k or K,	Cathode	
v or V,	Voltage	
i or I,	Current	
P,	Power	

$\left.\begin{array}{c} i \\ v \end{array}\right\}$ with subscripts $\left\{\begin{array}{c} c \\ b \\ e \end{array}\right\}$ or $\left\{\begin{array}{c} d \\ g \\ s \end{array}\right\}$ represents the instantaneous value of a varying component.

$\left.\begin{array}{c} i \\ v \end{array}\right\}$ with subscripts $\left\{\begin{array}{cc} C & D \\ B \text{ or } G \\ E & S \end{array}\right\}$ represents the instantaneous total value.

$\left.\begin{array}{c} I \\ V \end{array}\right\}$ with subscripts $\left\{\begin{array}{cc} c & d \\ b \text{ or } g \\ e & s \end{array}\right\}$ represents the r.m.s. value of a varying component.

$\left.\begin{array}{c} I \\ V \end{array}\right\}$ with subscripts $\left\{\begin{array}{cc} C & D \\ B \text{ or } G \\ E & S \end{array}\right\}$ represents the d.c. or no-signal value.

Maximum and minimum values are indicated by the use of subscripts (max) and (min).

Electrical parameters

	Device	Associated circuit	Basic units	
Resistance	r	R		
Reactance	x	X	Ω	Ohms
Impedance	z	Z		
Admittance	y	Y		
Conductance	g	G	S	Siemens
Inductance	l	L	H	Henrys
Capacitance	c	C	F	Farads

Double subscripts

Where capital V has two capital subscripts, this represents a voltage supply provided for the particular electrode of the device. For example, V_{CC} refers to a transistor collector supply voltage, and V_{DD} represents a voltage supply provided specifically for the drain of a FET. Where the two subscripts are different, the first subscript denotes the terminal at which the voltage is measured and the second subscript denotes the reference terminal. For example, V_{BE} represents the d.c. or no-signal voltage measured at the base, with respect to the emitter of a bipolar transistor. Other parameters may also be denoted by the use of double subscripts, as with c_{gs} the capacitance which exists between the gate and source of a FET, or $r_{bb'}$ the resistance between the actual base point and the base connector.

Matrix notation

The first subscript in the matrix notation for semiconductor devices identifies the element of the four-pole matrix. Thus,

> i input
> o output
> f forward
> r reverse

A second subscript is used to identify the circuit configuration:

> e common emitter s common source
> b common base g common gate
> c common collector d common drain

For example, using h parameters for bipolar transistors,

$h_{ib}(h_{11})$ — The small signal value of input impedance with the output short circuited to alternating current.

$h_{rb}(h_{12})$ — The small signal value of reverse voltage transfer ratio with the output voltage held constant.

$\left.\begin{array}{l}h_{fb}(h_{21},\,\alpha)\\ h_{fe}(h'_{21},\,\beta)\end{array}\right\}$ — The small signal forward current transfer ratio with the output short circuited to alternating current.

$\left.\begin{array}{l}h_{ob}(h_{22})\\ h_{oe}(h'_{22})\end{array}\right\}$ — The small signal value of the output admittance with the input open circuited to alternating current.

Static values of parameters are indicated by capital subscripts, e.g.

h_{FE} — The static value of the forward current transfer ratio, with the output voltage held constant.

Similarly, using y parameters for field effect transistors,

$y_{fs}(y_{21},\,g_m)$ — The small signal forward transfer conductance with the output voltage held at zero.

$y_{os}(y_{22},\,1/r_d)$ — The small signal value of output conductance with the input voltage held at zero.

Additional symbols

f_α }
f_β }   The frequency at which the magnitude of the parameter indicated by the subscript is 0.707 of the low-frequency value.

f_1 The frequency at which the modulus of h_{fe} equals unity.

I_H Holding current for an SCR.

I_{CBO} }
I_{CEO} }   Collector current when the collector is biased in the reverse direction with respect to the reference terminal and the other terminal is short circuited.

V_{BO} Break over voltage of an SCR.

V_{BR} Breakdown voltage.

$V_{(BR)CBO}$ }
$V_{(BR)CEO}$ }   The breakdown voltage between the terminal indicated by the first subscript when it is biased in the reverse direction with respect to the reference terminal and the other terminal is short circuited.

$V_{CE(sat)}$ Collector-to-emitter saturation voltage.

$V_{CE(knee)}$ Collector knee voltage.

P_{tot} Total power dissipated within a device.

T_{amb} Ambient temperature.

T_j Junction temperature.

r_{bb}, r_{be}, r_{bc}
r_{ce}, C_{be}, C_{bc}
C_{ce}, g_m }   Components of the small signal hybrid-equivalent circuit.

g_m is also used as the mutual conductance of a thermionic valve.

r_a the anode resistance of a thermionic valve.

μ the amplification factor of a valve.

r_d the drain resistance of a FET, ($= 1/y_{os}$).

r_g the gate resistance of a FET, ($= 1/y_{is}$).

Bibliography

1. Shockley, W., *Bell Syst. Tech. J.* **28**, July 1949.
2. *The Hot Carrier Diode,* Hewlett Packard Application Note 907.
3. Mazda, F. F., *The Components of Computers–Optical Devices,* Elect. Components, London, Feb. 1973.
4. Gooch, C. H., *Electroluminescent Diode Displays,* Elect. Components, London, June 1972.
5. *Microwave Applications of Semiconductors,* Proc. Joint I.E.R.E.–I.E.E. Symp., London, July 1965.
6. *Applications of PIN Diodes,* Hewlett Packard Application Note 922.
7. Day, D. B., *J. Brit. Inst. Radio Engrs,* **21**, 3, London, Mar. 1961.
8. Penfield, P. and Rafuse, R. P., *Varactor Diode; Varactor Applications,* M.I.T. Press, Cambridge, Mass., 1962.
9. *Harmonic Generation using Step Recovery Diodes,* Hewlett Packard Application Note 920.
10. Hosking, M. W., *The Realm of Microwaves,* Wireless World, London, Feb. 1973.
11. Carroll, J. F., *Impatt, Trapatt, Gunn and l.s.a. Devices,* Arnold, London, 1970.
12. Thompson, P. A. and Bateson, J., *J. Brit. Inst. Radio Engrs,* **22**, 1, London, July 1961.
13. Shockley, W., *Holes and Electrons in Semiconductors,* Van Nostrand, New York, 1950.
14. James, J. R. and Bradley, D. J., *Electron. Technol.* **38**, Mar. 1961.
15. Ebers, J. J. and Moll, J. L., *Proc. Inst. Radio Engrs,* **42**, New York, Dec. 1954.
16. Ryder-Smith, S. C., *Electron. Technol.* **38**, Oct. 1961.
17. Chandi, S. K., *Trans. Inst. Radio Engrs,* C.T. 4, New York, Sept. 1957.
18. Shea, R. F., *Transistor Circuit Engineering,* Wiley, New York, 1958.
19. *Silicon Controlled Rectifiers,* A.E.I. Application Report 4450–205.
20. *Thyristors and Rectifiers,* R.C.A. Data book SSD-206.
21. Mazda, F. F., *Controlled Rectification,* Elect. Components, London, Jan. 1971.
22. *Heatsinks for Semiconductor Rectifiers and Thyristors,* S.T.C. Publication MF/187 X.
23. Bisson, D. K. and Dyer, R.F., *Trans. A.I.E.E. Comm. and Elect.,* May 1959.
24. *Transistor Manual,* General Electric Inc., 1964.
25. Crawford, R. H. and Dean, R. T., *The How and Why of Unijunction Transistors,* Texas Inst. Publication.
26. Watson, J., *An Introduction to Field Effect Transistors,* Siliconix Publication.
27. Sevin, L. J., *Field Effect Transistors,* Texas Inst. Publication, April 1963.
28. Sparkes, J. J., *The Radio and Elect. Engr.* **43**, London, Jan. 1973.
29. Brothers, J. S., *The Radio and Elect. Engr.* **43**, London, Jan. 1973.
30. Dean, K. J., *The Radio and Elect. Engr.* **43**, London, Jan. 1973.
31. Gosling, W., *The Radio and Elect. Engr.* **43**, London, Jan. 1973.
32. Boyle, W. S. and Smith, G. E., *Bell Syst. Tech. J.* **49**, 1970.

33. Beynon, J. D., *Electronics and Power*, I.E.E., London, May 1973.
34. Amelio, G. F., *The Impact of Large CCD Arrays*, Elect. Components, London, Feb. 1975.
35. *International Conference on the Technology and Applications of Charge-coupled Devices*, I.E.E., R.R.E., Edinburgh University, Sept. 1974.
36. Valley, G. E. and Wallman, H., *Vacuum Tube Amplifiers*, McGraw-Hill, New York, 1948.
37. Sturley, K. R., *Radio Receiver Design*, Chapman & Hall, London, 1947.
38. Langford-Smith, F., *Radio Designers Handbook*, Iliffe, London, 1953.
39. Joyce, M. V. and Clarke, K. K., *Transistor Circuit Analysis*, Addison-Wesley, New York, 1961.
40. Korne, G. A. and Korne, T. M., *Electronic Analog Computers*, McGraw-Hill, New York, 1952.
41. Paul, R. J., *Fundamental Analogue Techniques*, Blackie, London, 1964.
42. Hyndman, D. E., *Analog and Hybrid Computing*, Pergamon, Oxford, 1970.
43. Lynch, W. A., *Proc. Inst. Radio Engrs*, **39**, New York, Sept. 1951.
44. Benson, F. A., *Electron. Engng*, **24**, Sept. 1952.
45. Walker, D. E., *Electron. Engng*, **34**, June 1962.
46. Brunetti, C., *Proc. Inst. Radio Engrs*, **27**, New York, 1929.
47. Sommers, H. S., *Proc. Inst. Radio Engrs*, **47**, New York, 1959.
48. van der Pol, B., *Phil. Mag.* **2**, 1926.
49. Abraham, H. and Bloch, E., *Ann. der Physik*, **12**, 1919.
50. Millman, J. and Taub, H., *Pulse and Digital Circuits*, McGraw-Hill, New York, 1956.
51. Eccles, W. H. and Jordan, F. W., *Radio Rev.* **1**, London, 1919.
52. Chance, B., *Waveforms*, McGraw-Hill, New York, 1948.
53. Neeteson, P. A., *Junction Transistors in Pulse Circuits*, Philips Tech. Lib., Eindhoven, 1959.
54. Beaufoy, R., *J. Inst. Elect. Engrs*, May 1959.
55. Dummer, G. W., *Fixed Resistors*, Pitman, London, 1956.
56. Bardsley, M. and Dyson, A. F., *The Radio and Elect. Engr.*, **44**, 4, London, April 1974.
57. Dummer, G. W., *Fixed Capacitors*, Pitman, London, 1956.
58. *Electron. Prod. Mag.*, U.T.P. New York, Dec. 1968.
59. Pearlston, C. B., *Trans. Inst. Radio Engrs*, R.F. **14**, New York, Oct. 1962.
60. Jaeger, J. C., *An Introduction to the Laplace Transformation with Engineering Applications*, Methuen, London, 1949.
61. van Valkenburg, M. E., *Introduction to Modern Network Synthesis*, Wiley, New York, 1959.
62. Weinburg, I., *Network Analysis and Synthesis*, McGraw-Hill, New York, 1962.
63. Zimmerman, H. J. and Mason, S. J., *Electronic Circuitry Theory*, Wiley, New York, 1959.
64. Langmuir, I., *Phys. Rev.* **2**, 1913.
65. Valley, G. E. and Wallman, H., *Vacuum Tube Amplifiers*, McGraw-Hill, New York, 1948.
66. Miller, J. M., *Sci. Pap. U.S. Bur. Stand.*, No 351, 1919.

Index